Computing for architects

R. A. Reynolds BSc, AMBCS

Butterworths

London Boston Durban Singapore Sydney Toronto Wellington

This book is sold subject to the Standard Conditions of Sale of
Net Books and may not be resold in the UK below the net price
given by the Publishers in their current price list.

First published 1987

© **Butterworth & Co (Publishers) Ltd, 1987**

British Library Cataloguing in Publication Data

Reynolds, R. A.
　　Computing for architects
　　1. Architecture—Data processing
　　I. Title
721′.028′5　　　NA2728
ISBN 0-408-00800-8

Library of Congress Cataloging in Publication Data

Reynolds, R. A.
　　Computing for architects.

　　Bibliography: p.
　　Includes index.
　　1. Architecture—Data processing.　I. Title.
NA2728.R49　1987　　　720′.28.5　　　86-34305
ISBN 0-408-00800-8

Photoset by Butterworths Litho Preparation Department
Printed in Great Britain at the University Press, Cambridge

Preface

This book aims to help with the introduction and use of computers in architectural offices. It is the result of seventeen years' experience of using computers in firms in private practice and is intended to be more of a practical guide than a textbook. I hope to show where computers can help and how they can be applied and also hope to point out the worst pitfalls, into most of which I have fallen myself at some time.

The text does not go deeply into the details of such matters as binary arithmetic and programming languages. A knowledge of these is interesting but no more essential to the use of a computer than a knowledge of internal combustion theory is to driving a car. There is now a great deal of good software available, and the working architect will gain most by being aware of the principles involved and so be able to use it intelligently.

I have attempted to avoid using jargon and technical terms where possible, but just as it is difficult to describe building without using terms like 'purlin', 'soffit' or 'DPC', so it is difficult to describe a new discipline without using its language. However, I have tried to explain the meaning of new terms. I believe that the effort of becoming familiar with the computer will amply repay architects, who will come to find it one of the most powerful tools at their disposal.

Tony Reynolds

Acknowledgements

I began my professional career in computing with the firm of Cusdin, Burden and Howitt. The partners of this firm have supported the application of computers to architectural problems since 1964. I wish to thank the past and present partners for their persistence and foresight in sponsoring these developments over many years.

I am currently employed by the EPR Partnership. The partners of this firm have made a considerable investment in computer-aided design equipment and are also supporting developments in hardware and software. I wish to thank the partners for their encouragement and backing.

Finally, I would like to thank Miss Linsey Mendes for her skill in deciphering my handwriting and typing the manuscript.

Note. Computer techniques were used throughout the production of this book. The text was typed into a word-processing system, and the spelling checked automatically. The word-processor was linked into an automated typesetting machine which set up the print.

All the diagrams in this book were produced by computers. Except where otherwise acknowledged, these have been prepared by the author through the use of a computer-aided draughting system.

Contents

Chapter 1

Past and present

The attitudes of architects towards computers have undergone very sharp changes over the years, rather than the continuous development that might have been expected. Great enthusiasm in the early years was followed by a period of disillusionment when it was found that the capabilities of computers did not even begin to match the claims made for them. During this latter period work was carried on by a fairly small number of enthusiasts, and their developments, coupled with the continuous fall in the cost of machinery, led in time to a guarded acceptance among architects that computers could be of use in certain circumstances. Then a few years ago, with the introduction of microcomputers, the cost of computers dropped so dramatically that they could be afforded by almost everybody. Architects, in common with many other people, bought these machines for their personal use and there has been another surge of enthusiasm as attempts are made to apply the techniques of computing to the problems of the profession.

Computers were first introduced in the 1950s and were used initially for scientific calculations and for straightforward large-scale business applications such as payroll production. It was not long, however, before their capabilities were being explored in every field, including that of architecture. In 1963, Ivan Sutherland at the Massachusetts Institute of Technology announced the famous SKETCHPAD system[1]. This used a screen connected to a computer and allowed the user to produce line diagrams on the screen by the use of a special pen and a keyboard. Arcs, circles and straight lines could be produced with little effort, and repeated drawing elements and symbols could be defined once and thereafter called up and positioned as many times as necessary.

Three years later, William Newman at Imperial College in London developed a similar system for specifically architectural applications[2]. Standard modular building elements such as wall sections, windows and floor slabs were stored in the computer and the user was able to select from the range and assemble a plan on the screen. When the plan was finalised, the computer was able to produce a list of room areas automatically, compile a schedule of the materials used, calculate the heat loss from the structure and assess the natural and artificial lighting levels.

The same period saw the rise of a number of design theoreticians. These researchers attempted to find a logical basis for design and so devise systematic rules from which designs could be created. The methods they used often required the use of computers, to reduce to manageable form large tables of the interaction of each activity area with all other areas, for example, or to apply complex mathematical optimisation techniques. One of the best known of these workers was

1

Christopher Alexander, a mathematician and architect, who published a book[3] and many articles setting out his ideas which attracted much attention at the time. It was natural to feel that such techniques would soon become universal and lead to better designs.

In the mid-1960s the world economy was expanding. There was a building boom and architects were finding it difficult to keep up with their workload. Labour was expensive and hard to find. There was therefore both the incentive and the finance to develop solutions that would reduce the dependence on manual effort.

The other members of the construction team were also finding computers invaluable. The structural engineer and other consultants were able to have laborious calculations done very quickly and cheaply and so were able to give more attention to general principles and to the testing of alternative solutions. At the other end of the construction process, the larger contractors were saving important sums by using computers to schedule the optimum use of men and machinery on site and to organise the ordering of materials at the correct times.

It was against this background that the enthusiasm developed. The architectural profession has always shown a readiness to try new ideas, although whether this is a laudable tendency towards open-mindness, or a less laudable love of novelty is open to question, and the late 1960s and early 1970s saw a boom in attempts to use the computer. In that period not a year went by without several conferences on computer-aided architectural design being held[4,5], a number of books and hundreds of articles were published[6], and many working parties were set up to investigate and report on different aspects of the use of computer in the construction industry[7]. Researchers all over the world were writing programs and attempting to apply them to real-life building projects.

Most of the work done had little long-term impact, but some large-scale projects funded by government bodies were brought into operation. The architectural department of the UK's West Sussex County Council provided one instance. It developed an almost complete design system using computer aids applied to the industrialised building system SCOLA. The architects worked at a screen rather than on drawing boards and by 1968 they were able to follow the entire design process through. From feeding in the basic activity relationships they were able to define and position spaces, to correct for the Building Regulations and optimise for economic performance, heating requirements and many other things, and finally to produce priced and comprehensive bills of materials and production drawings on a computer-driven plotter[8].

Another instance of the new approach was given by the UK's Department of Health and Social Security (DHSS). The DHSS has always tried to help the architect in the design of medical buildings and has published comprehensive checklists and guidelines[9]. In 1969 the DHSS introduced an integrated hospital-design concept named HARNESS[10]. HARNESS aimed to provide a large measure of standardisation while at the same time retaining flexibility. Although the use of the computer was not essential to the system, it was considered to be an important aid to its viability. Programs were developed to evaluate the manual design of individual departments in terms of circulation efficiency, heat loss and gain, site usage and capital cost. Another set of programs automatically assembled the departments around a service spine in the optimum manner, subject to the constraints of the brief. There were also programs to produce production documentation, including drawings.

There were many other projects, mostly on a small scale and applied by private firms to individual buildings, but within a few years it became obvious that there was not going to be a revolution. When architects attempted to use computers for themselves they found that it was a full-time task with few rewards. A lot of new techniques had to be learned and problems solved, most of them in order to set to grips with the machine and circumvent the restrictions imposed by the relatively slow and small computers available at that time. There were not many programs on offer and those that were available mainly concentrated on problems that required a lot of calculations, such as beam design, daylight-factor analysis and heat loss evaluation: problems that were on the border of the architect's interest. The central problems that occupied most of an architect's time were not tackled at all. Further, those programs usually required large amounts of data collection and preparation and gave results that were not at all commensurate with the time and effort put in.

The much-vaunted draughting systems that were to replace drawing boards with screens turned out to cost more than the salaries of dozens of draughtsmen. This virtually ruled out their use in any but government-supported organisations, and even in such organisations it was found that draughting speeds were little, if at all higher than the manual equivalent.

The building design programs were found to produce very naive designs. Normally they would operate by attempting to optimise a single factor, circulation cost being the most popular choice. They did not take account of the thousands of other factors that must be considered in producing a workable design, and in a short time these programs were only being used in experimental applications. The general feeling grew at that time that an architect's work was simply too complex and too intuitive to be aided by computers.

By the mid-1970s this attitude and a number of other factors combined to make disillusionment with computers complete. The oil crisis of 1974 changed the world economic climate rapidly for the worse. The construction workload fell dramatically and architects found themselves with insufficient work and plenty of people about to do what work there was. Public opinion was turning against industrialised building systems. It was found that such buildings cost almost as much as conventional buildings but were inferior in a number of ways and were disliked by the people who had to live in them. Very little new industrialised building was erected, and industrialised building is the most efficient in computer usage as it involves choices from a limited range of components, details of which can be given to the computer in advance. Conventional building methods, on the other hand, can involve an almost unlimited number of components.

Some of the ambitious computer projects quietly closed down. West Sussex County Council Architect's Department went back to manual methods in 1974 after it was decided that the reduced use of the SCOLA system did not justify the cost of using the computer. The DHSS abandoned the HARNESS system in 1975 after it was found that the buildings produced were too expensive in the new financial climate, and had incurred development costs estimated by one source to exceed £3 million[11]. Most architects who had tried to use the computer gave up, deciding that the results did not justify the effort or the cost.

In the middle and late 1970s there was little general interest in computers in the profession, but despite this a fair amount of research and development work was carried out. This effort was much more realistic and hard-headed in terms of what could be achieved and what was cost-effective.

A number of organisations were active in research and development, but a few can be singled out as among the more influential. In the government-supported field, the Computer-Aided Design Centre was set up in 1969 by the UK's Ministry of Technology. Programs applicable to the construction industry became available from the Centre a few years later. The concepts they developed in this period still underlie a number of popular systems, including those marketed by the Centre itself, which has become a private company under the name of the CADCentre Ltd.

In the academic field, the Architecture and Building Aids Computer Unit, Strathclyde (ABACUS) wrote a wide range of relatively small programs attacking specific problems. Among private architectural practices, one of the most impressive achievements was that of Perry, Dean and Stewart of Boston, Massachusetts. This firm developed one of the first practical draughting and design systems, called ARK/2, in 1970[12] and subsequently marketed it with some success. In the commercial field, the software house of Applied Research of Cambridge Ltd was set up in 1969 by a group of architects and began the development of various programs, most notably OXSYS, an integrated building and design system[13].

Such work as the above was encouraged by the fall in the cost of computer power. New techniques were allowing the equivalent of thousands or tens of thousands of separate components to be contained within an element no larger than a fingernail. An electronic calculator in 1968 was the size of a typewriter and cost the equivalent of a month's wages; by 1975 it could fit into a pocket and cost a few hours wages. A similar scaling-down was going on throughout the field. Minicomputers arrived in the early 1970s and for the first time ordinary architectural practices were able to afford their own machine.

As well as being cheaper, minicomputers were easier to use. The manufacturers had had time to develop the programs to control the computers and they were provided with standard utilities that removed a lot of technical responsibilities from the user.

The power of the computers at this time was also expanding rapidly. In 1968 there was only one computer available in the UK that offered a public service by telephone communication. This computer could only accept user programs of up to 250 instructions and data files up to 6,144 characters. By the mid-1970s there were many computers of this type and their limitations of use were not normally restrictive. This allowed programs not only to solve the problems, but also to be more 'user-friendly' and therefore easier to use by less technical people.

Towards the late 1970s, continual program development and reduced circuitry cost had progressed far enough for a number of the larger practices to be able to justify installing a computer-aided draughting system.

It was around 1980 that the first microcomputers arrived. Suddenly, computers were only a tenth of the cost of the cheapest machine previously available. Many people bought one and with mass sales the costs fell even further. In response, there was an abrupt flowering of interest in all fields, including architecture. Thousands of architects learned to use these machines and began developing their own programs and buying and exchanging programs with other enthusiasts.

The current situation is that we are in a very exciting period of fast development both in machinery and in programming. Microcomputers are now so widespread that one household in two in the UK has one[14]. There has been, however, a change in emphasis and in scale from previous development. Because most microcomputers have small memories and low speeds, the programs they support must apply themselves to fairly specific problems. Even the larger 'business' microcomputers do not have the power to handle some problems adequately, notably the searching of large amounts of data or the production of complex drawings, so in some areas minicomputers are still favoured.

Microcomputers and their programs are now so cheap that questions of cost-justification are no longer relevant. The range of programs available for architectural users is fairly wide, and is growing fast. A recent survey found that 67% of architectural firms in the UK own a computer[15] and most architects now accept that computers can aid their work, although their expectations are more realistic than they were in the 1960s. There is today a widespread eagerness to learn how to use computers and how to get the best from them.

Chapter 2

Overview

Benefits of using computers

Over the years, attempts have been made to use computers to aid virtually every task in the construction process. The programs currently available that are of interest to the architect cover a wide range. They include programs which provide assistance at a very general level, such as those that estimate costings from the rough massing of the structure, and programs which help with very precise and well-defined activities, such as predicting if condensation is likely to occur within a given construction.

Most of the advantages of computers derive from their being able to carry out long sequences of well-defined operations at high speed. The operations themselves are elementary, but millions of them can be executed in a single second and give the net impression of sophisticated behaviour.

At the most straightforward level, this high speed can be harnessed to carry out complex calculations in a short time with great accuracy. For example, the calculation of daylighting levels is very laborious if done manually with any precision. It involves the consideration of the colour and texture of internal surfaces and of the amount and nature of external obstructions as well as the shape and size of the window itself.

In practice, such calculations are rarely done at all by hand. The architect will where necessary perform rough checks with daylight protractors or tables and then add a sizeable 'safety factor' to the window sizes to ensure that the occupants get a reasonable amount of natural light. This can in turn lead to excessive solar gain and glare and to higher heat loss than necessary. The use of computers can therefore give a better and more economical design.

Even when done with accuracy, the complexity of calculations can mean that they are only undertaken when the design is finalised. For instance, the exact calculation of heat loss from a structure and from individual rooms will normally be done by the services consultant at the end of the design stage. These values will then be used to determine the heating provision. However, if the architect has access to quick and cheap calculations of this type, different building layouts can be tested at an early stage in the design process and very possibly the final installed heating cost will be reduced.

The high speed of calculation can also be used to shorten the design period. This has become very important because clients are demanding much shorter construction times than previously. Whereas building time used to be measured in years, it must now be measured in months. The reasons for this are mainly financial: the very high level of investment in large buildings requires fast construction, especially in periods of high inflation, if the project is to be economically viable.

The use of computers can speed up many everyday tasks in an architect's office. A simple example might be the production of door and window schedules. At present, these can take weeks of very tedious effort, but computer programs exist that can help with the production and revision of these, or even in some cases generate them completely automatically.

As buildings have become larger and the time for design has shortened, architectural practices have reacted by increasing the size of the design teams. This process is one of diminishing returns as the employment of more staff causes serious financial losses in the extra administration and co-ordination necessary. When the team becomes large, a great deal of time is lost in communication between individuals. By increasing productivity, the computer can keep the team small and cohesive and so makes further gains by reducing the overheads of control and communications.

Computers can speed up most parts of the design process, but the most important savings will come in the faster generation of production documentation. This documentation, including working drawings and details, specifications, schedules, Architects' Instructions, Bills of Quantities and tender documents, takes as long and consumes twice as many man-hours as the design itself. The current distribution of an architect's time as reflected by the recommended fee scales of the American Institute of Architects[16] and of the Royal Institute of British Architects[17] is illustrated in *Figure 2.1*.

All the indications are that paperwork can only increase and this will further encroach on the time needed for design. Fortunately, it is precisely in the well-defined and less intuitive areas that the computer is at its best.

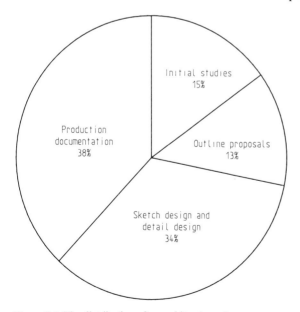

Figure 2.1 The distribution of an architect's work

Another way in which computers can help is by giving the architect easy access to stores of information and knowledge. Buildings are not only getting bigger, but are also becoming more complex. They have to provide an ever-widening range of services, and the rise in user expectations has forced the architect to pay more attention to such matters as better lighting, air-conditioning and thermal comfort. Tighter legislation requires the architect to conform to many rules on construction, energy saving, fire precautions etc. Advances in technology provide thousands of new materials and techniques every year, many of which can be of great use in construction, but unless designers are to spend all their time reading they cannot hope to be aware even of the existence of most of them.

The computer can be of help here in two ways. First, a computer program can incorporate within itself expert knowledge and techniques. Thus computer programs exist to suggest viable lift provisions, to check designs for conformity with building regulations, to size and lay out air-conditioning ducts etc. Many such programs are available.

The second way a computer can help to cope with complexity is through its ability to hold, scan and restructure large amounts of information. Databases exist for technical articles, legal references, products, components, etc. For example the architect could scan for flooring materials and list out products of a suitable appearance, texture, strength and cost for a particular job. This might replace days of research or provide an optimum result when otherwise the designer would specify a product that has been used before, just because of its familiarity.

One consequence of the increased complexity of buildings has been the growth in the use of specialist consultants. This fragmentation of the design team has been happening for a long time. In 1842, Pugin is recorded as having worked on 25 sizeable schemes aided by only one assistant[18]. By the 1920s a consultant engineer was normally used for larger projects and in the UK the quantity surveyor was formally established to prepare bills of quantities for costing and ordering.

Today, a large project may use an architect, a structural engineer, a quantity surveyor or cost advisor, a services engineer, an environmental consultant, a landscape architect, an interior designer and even specialist designers who completely take over the design of highly serviced areas such as kitchens and laboratories. *Figure 2.2* illustrates this trend and is based on the average fees paid to the participants.

Architects are becoming less concerned with design and more concerned with the administration and co-ordination of all these different disciplines; a task for which they have not even been formally trained. The almost inevitable result has been the appearance of building consultancy firms which deal directly with the client and control the whole design and construction process, merely employing architects for the design work.

Computers offer the possibility of reversing this trend. They can easily deal with the formulae and rules that most of the other disciplines use and can have access to large amounts of specialist information. While it may not be possible entirely to dispense with outside consultants, their use can be greatly reduced. With a knowledge of the general principles involved, an architect can test a hypothetical design and modify it in the light of the results until the best solution or the best compromise emerges. This will save time because of the reduced overheads of administration and communication. The architect need no longer send his drawings to the consultant for checking and comments. It will also typically result in a better design as the architect has more insight and more control over the solution.

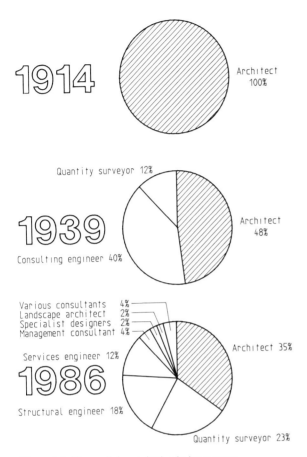

Figure 2.2 The participants in the design process

A less important, but still valuable, gain from computer methods is that information can be recycled much more easily for use in later jobs. At present, little feedback takes place because the time and manpower necessary to correlate and index the information are not available and therefore reference to earlier designs is difficult. However, once fed with information, the computer can store it and at a later time recover it and integrate it with other data to provide a base for a new job.

An example of this at an elementary level might be a constructional specification document. Standard paragraphs detailing exactly how the contractor should carry out his work can be stored in the computer and recalled as required. Paragraphs necessary to a particular job can be added and a finished document printed out very quickly. The new paragraphs can then be added to the information store for use on future jobs.

At a more complex level, the use of a computer draughting system can mean that a layout for an entire area, such as a hospital ward, is retrievable. If that layout should be thought suitable in the design of a later hospital, the computer can be instructed to perform a 'cut and paste' operation to make the ward part of a new drawing.

Drawbacks of using computers

Computers are not an unmixed blessing. They are unsuited, or not well suited, to certain kinds of problems and they inevitably introduce difficulties of their own. The nature of the problem, or of the building or its scale, might preclude the efficient use of computers. Computers may impose extra and unfamiliar duties upon the architect which may disrupt traditional ways of working.

Obviously, a problem whose solution depends wholly or partly on subjective judgements is not normally suited to computer solutions. For example, aesthetic evaluations or the considerations of human behaviour and relationships cannot adequately be dealt with by machine. This can be circumvented to a certain extent. For instance, there has been an interesting project which used the machine to generate a series of random arrangements of housing units on a hilltop site within the constraints of the brief[19]. This approach could provide useful ideas for the designer, but computers can only work if given a set of systematic rules, and such rules do not exist to define human emotions or judgements.

It may also be inefficient to use the computer if the building itself does not give the computer enough scope. Because much of the power of computers lies in their ability to apply the same process again and again, the building must contain a reasonable amount of repetition. If this is not the case, the information or process used in one situation may not be applicable to any other, so giving no savings in time.

A simple example might be given by a door schedule. If almost every door in a building is different, it will take longer and cost more to type in the details than it would do to produce a conventional schedule on squared paper. However, if, say, a range of ten doors has been specified for the building, then it would be quicker to produce a computer schedule than the manual equivalent because it takes little time to input the details of the ten types and the computer can associate these with a compact list of door numbers. A similar situation exists for most other computer aids. Thus a building that is completely unrationalised cannot be handled efficiently by computer.

A building may be so small that it cannot give much scope for repetition. Such a building will also be much more easily comprehensible and controllable by a small team. Therefore there are few communication problems and little possibility of extensive checks being necessary. Again, in these circumstances the computer can offer little help; in fact the extra overhead of using the machine may well actually slow the design process.

Surprisingly, it is also possible for a building to have too high a degree of repetition. If a building has many floors, each of which is identical, as might be the case in a hotel above the ground floor level, then a single plan of a typical floor will serve for many. Thus, although the building itself may be large, as a design problem it is equivalent to a small building.

A building can also be of a type that is difficult to handle by computer because it is unusual. In order to make them simple to operate, many computer programs make assumptions about buildings which are not always valid. At an elementary level, a program to check the amount of daylight coming through a window will assume that the window is a rectangle. If it did not, a lot more data would be necessary and the program would have to be far more complicated than it is merely to handle the very rare case of a non-rectangular window. This means that the architect who wants to specify a parallelogram-shaped window on a staircase, or a circular shape as in the traditional 'rose' window, will not be able to use the computer to check the natural lighting.

At a more important level, some computer-aided draughting programs are difficult to use when the walls in the building do not run parallel to one of the edges of the paper. This would preclude the use of such a program on a building that is circular in plan or a building that curves to fit the shape of a road or a seafront. Equivalent assumptions exist in many other types of program.

A problem in the use of a computer that troubles many architects is its tendency to mould the design. This is of course a problem with any tool: the tee-square and set-square undoubtedly encourage the architect to draw a certain sort of building. The computer can encourage a move towards more standardisation and repetition than would otherwise be necessary, and this may be detrimental to the design. Even if a specific situation calls for a particular solution, the architect might be tempted to use the nearest equivalent from a built-in library of standards rather than to go to the trouble of defining a special solution that may never be used again.

This may not be entirely a bad thing if it discourages an undisciplined approach to the design problem. It is also true that the growth in size and complexity of buildings has already forced the profession to make a lot of concessions to rationalisation. Most buildings of any size are now planned on a modular grid and efforts are always being made, with varying degrees of success, to use standard ranges of components, standard working details and even standard layouts for certain complex types of activity areas.

The computer can cause dislocation in the office because it forces the architect to adjust his method of working. One aspect of this is that the architect must work in a much more systematic way than formerly. As the computer requires precise instructions and completely accurate data there is a multitude of small but intricate tasks to perform before the machine can begin to work. The data must be prepared, the program accessed, the inevitable errors in the data corrected and a resubmission made, etc. It is therefore much more efficient to process large amounts of data on relatively few occasions. This principle might cause problems if the architect has been used to alternating between different aspects of a job. Manually, such alternation loses little time in the end and may even be a better way of working as it increases variety. When using computer methods, such freedom is restricted.

Personal problems may also occur because some architects will be unwilling or unable to use the computer. Often some members of staff will not want to change working methods which they have become used to and have found effective over the years. Also often some members of staff will simply find that they are unable to grasp the principles on which the machine works and will never be able to get the best out of it, however enthusiastic they are. Older architects are especially prone to both these problems, but they are found to some extent in all age groups.

Because of the difficulties involved in using computers, not least of these being the slow typing speed of most architects, some firms prefer to employ a specialist intermediary to actually perform the operations. This can increase efficiency: one study has shown that the use of a trained intermediary can cut computer usage to one-sixth of that incurred by casual users[20]. The problem with this is that the architect will often feel at a distance from the computer and so will never come to understand exactly what it can do and how it does it.

In the worst cases there will arise what has been called the 'whizz-kid barrier' where the computer user talks in terms completely different from those of the architect and mutual incomprehension obstructs the workflow. Fortunately, microcomputers are so cheap that intermediaries are not usually necessary. However, it can still be problem if the office chooses to install a more powerful and expensive machine which must be used intensively to be viable.

The quality of life can also suffer. One of the most persistent computer myths is that computers will relieve the architect of the tedium of repetitive tasks. To a large extent the opposite is true. The machine must have data before it can work and tasks such as scheduling will require the collection and checking of large amounts of information. Eventually, of course, time will be saved, but it will still be at the price of a good deal of boring and exacting effort. The point has been neatly put by one worker in the field, 'It has been frequently argued that computerised equipment could free man from the soul-destroying, routine, back-breaking tasks to engage in more creative work. Anybody who looks at a highly automated factory must simply question whether this is, in fact, so.'[21].

It may also be necessary in some cases to introduce shift working where computers are used. This can occur for several reasons. One reason is of course, that where the machinery is very expensive it must be used intensively to be cost-effective. This is still often the case in the computer-aided draughting field where the computers themselves are more powerful and also need expensive ancillary devices such as screens and plotters. On the other hand, shift working may also be necessary with microcomputers because of their low speed. Working through large amounts of data can take hours, and if it is not to hold up the rest of the design team it may be necessary to work at night. Further, if a job is behind schedule it is possible to flood it with temporary staff when working manually, while it is not usually practicable to acquire extra or more powerful computers at short notice. In this case, twenty-four hour working on the computer may be the only answer.

Another problem is that computers impose a high demand on the user. When a computer is used in a question-and-answer mode it does tend to drive the person using it. Because the computer works and reacts so much faster than humans it is usually waiting for a response. This paces the user at a higher rate than normal and the difficulties of maintaining this rate can lead to stress. The UK Health and Safety Executive recommends short, frequent pauses from work, preferably away from the machine in order to avoid fatigue[22].

I have made some observations of a group of operators using a computer-aided draughting system and have found that in terms of the number of responses made their performance was highest in the second hour, then dropped to less than half of that value after the fourth hour[23]. *Figure 2.3* illustrates these findings.

Outline design

The outline design stage is the least developed of all areas in the construction process as far as computer aids are concerned. This is mainly because so much of early design work is subjective. However, some programs do exist and in the right circumstances they can make the difference between a design that copes adequately with its requirements and one that does not.

Most programs in this field can be divided into the generative and the analytical. The generative programs actually produce a solution, with or without some human intervention, whereas the analytical programs test or comment on a given design which the user can then modify and re-test.

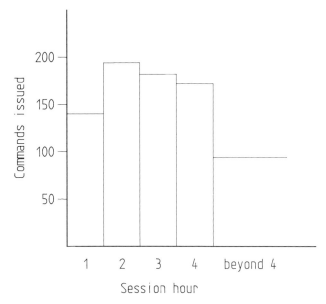

Figure 2.3 Operator performance on a computer-aided draughting system

The earliest examples of generative programs were those that produced plan layouts of rooms. There is less interest in these today, but some are still in use. Given room areas and their relationships to other rooms, the programs will arrange them into a plan shape. The criteria for doing this will be objective: the relationships given by the user will be the main basis for the solution, but a sophisticated program will make sure that all occupied rooms have natural light and that the building shape is an economical one from the point of view of construction and of heat loss.

More specialised forms of generative program are available to produce solutions in certain specific areas; for example, programs exist to design stairways. The programs contain the building regulations that lay down the permitted relationships of tread to riser for different types of building. Other regulations or advisory guidelines will specify the maximum number of steps before a landing must be inserted, the height of the stair rails and so on. The user need only give the floor-to-floor height and an indication of the type of building being constructed and the computer will produce a design. The more advanced programs will even draw out the section and plans and some details. Programs such as these can save a fair amount of reference and calculations.

Analytical programs can be of several types. Programs exist that can test a given design for fixed parameters. In practice, most of these are environmental, so for a given massing and orientation the heat loss can be calculated. Another example is programs that calculate cut-and-fill; that is, the amount of earth that must be removed or deposited before the foundations can be laid. On a sloping site this can be expensive, and the architect might test early solutions on this basis and use the results as one of the factors in the final decision.

Another type of analytical program is a simulation system. Simulation is the setting up of a mathematical model within the computer to mimic movements and processes. In architectural terms, most of the movements will be those of people through a building. When the model is constructed the designer can run it under different conditions and see what happens. It might be found, for example, that unacceptably large queues form at certain points or at certain times of the day. Alternatively some facilities might be under-used. These programs can be of use for a wide variety of buildings. They can, for instance, be used to test a supermarket for adequate but not wasteful provision of checkout tills, car-parking space, aisle widths, etc.

Figure 2.4 shows an output from a computer simulation predicting the fluctuations in the occupancy of delivery rooms in a maternity ward. The designer can actually watch a speeded-up animated display and observe the consequences of the design.

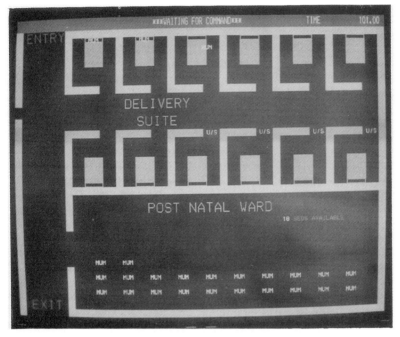

Figure 2.4 An animated simulation output (courtesy P-E Information Systems Ltd.)

Some simulation programs come packaged for specific situations so as to cut down on the amount of work the user has to do. A popular packaged system exists to model lift provision; the designer is queried on the number, size and speed of the lifts and must also give some information regarding the building. The program can then be used to test the effect of people arriving in different numbers and patterns on the various floors and will print out the average waiting times and queue lengths.

A third type of analytical program is the expert or knowledge-based system. An expert system is a program that can answer questions in a flexible way within a specialised area of expertise. The system has a database of information to draw on, which is normally relatively small, and which is structured in such a way as to make it possible for the program to follow a line of reasoning. Straight-forward examples of expert systems are those which have been written to answer legal queries. The programs have a knowledge of the relevant laws and will pose questions for the user to answer. They will finally print a judgement together with the relevant paragraphs of the laws.

More complex systems operate within a framework of probabilities. Thus the answer to a question may indicate that several possibilities are present at varying degrees of likelihood. Some programs of this type exist to comment on various types of building defects. In this case, a visible symptom, such as dampness, might be due to various causes with different probability. The program attempts to clarify this and identify the fault by asking further questions. These programs can be extremely effective. In a famous trial of the clinical diagnosis of blood infections, an expert system out-performed even very experienced practitioners[24].

Information handling

The computer's ability to store, retrieve and manipulate large amounts of information in a short time is very valuable in many ways. The amount of information needed to document a building is now very great, and the amount that should be researched when designing the building is immense.

Most large practices find it necessary to have a library and a full-time librarian, and even small offices will have several filing cabinets full of manufacturers' literature and articles from technical journals. The collating and updating of all this information is a slow and boring task which often falls behind schedule. It is clear that as more information is produced on every subject and as manufacturers extend their ranges, offices will become even more seriously overloaded.

In principle, the computer can go a long way towards reversing this situation. Physically, the information will occupy little space. It can also be updated very quickly and is self-collating, so that the data can be retrieved much more easily. The output can also often be used directly as part of the job documentation.

Databases can be set up by the architect for particular purposes within the office, and pre-prepared databases are also available which are created and updated by third parties. These latter cover a range of information. One very useful area is that of specifications. In the USA, the American Institute of Architects Service Corporation offers 'MASTERSPEC', which is a database of standard specification paragraphs. These can be selected as required, then using word processing techniques can be added to or modified to make up the complete specification. In the UK, the Royal Institute of British Architects offers a similar service in its National Building Specification.

Commercially-available databases exist that cover many subjects. Those of interest to architects include references to technical literature, a legal guide, a builders merchant's catalogue, current news in the construction industry and a software guide. Access to these databases may be through the public telephone network, or may be by the circulation of magnetic discs which the user puts into his own computer. Access costs are low and are in many cases free when the manufacturers whose products are listed pay for inclusion.

By using a generalised database program, the architect can set up his own databases. These can cover a wide range of subjects. One small, but useful, type of database might be that for client records. A large office might have many clients over a few years. A list of those clients with their associated data, including descriptions of the buildings they have commissioned, can be valuable. Contacts can be maintained more easily at a personal level, a mailing list can be produced, a list of buildings of a given type or size can be obtained to show potential clients, and so on.

Larger databases can be set up for particular jobs. Popular subjects include doors, windows, room finishes and fittings. Because of the computer's ability to repeat information, similar items at different locations can be tagged with the correct data more quickly than by manual methods. The printed result will have a much better appearance and most powerfully of all, the information can be scanned. This means that consistent changes can be made throughout. If one type of glass in the windows is changed for another type throughout, for example, the computer can be ordered to make the change. In certain situations it is also useful to use scanning to interrogate the database. In a services database it would be possible, say, to extract all rooms of certain type and check that a telephone outlet has been specified. *Figure 2.5* shows a page from a typical computer-produced schedule of equipment.

Computer-aided draughting

The ability of computers to draw is undoubtedly their most exciting aspect for architects, and one that gripped their imagination for a long time before it became economically viable. Clearly, it is this ability that gets right to the heart of an architects' work: the manipulation of spatial concepts. Since Ivan Sutherland's SKETCHPAD system, it has been hoped that spectacular results could be obtained

```
EAST CHEAM UNIVERSITY LIBRARY - FURNITURE AND EQUIPMENT
30 JULY 1986

ROOM
NUMBER  ROOM NAME                         CODE   QTY   DESCRIPTION


2212A   GOVERNMENT PUBLICATIONS STACKS    BIN11    2   BIN,WASTE PAPER
                                          BRD11    1   BOARD,NOTICE,PIN-UP,1000MM (W) 600MM (H)
                                          FIR11    4   FIRE-FIGHTING EQUIPMENT,CO2 EXTINGUISHER
                                          KIK11    3   KICKSTEP
                                          STA21   22   STACK,DOUBLE-SIDED,2300MM (H) WITH 200MM SHELVES


2212B   STACK ZONE, STATISTICS            BIN11    2   BIN,WASTE PAPER
                                          BRD11    1   BOARD,NOTICE,PIN-UP,1000MM (W) 600MM (H)
                                          STA21   44   STACK,DOUBLE-SIDED,2300MM (H) WITH 200MM SHELVES


2212C   PERIODICALS STACKS                BIN11    2   BIN,WASTE PAPER
                                          BRD11    1   BOARD,NOTICE,PIN-UP,1000MM (W) 600MM (H)
                                          KIK11    2   KICKSTEP
                                          STA27   28   STACK,DOUBLE-SIDED,2300MM (H) FOR PERIODICALS


2212D   READER ZONE                       CHA15  105   CHAIR,MEDIUM COMFORT
                                          REA32  105   READER PLACE,1000MM X 600MM

2301    GENERAL COLLECTION                BIN11    4   BIN,WASTE PAPER
                                          BRD11    1   BOARD,NOTICE,PIN-UP,1000MM (W) 600MM (H)
                                          CAS11    1   CASH BOX
                                          CHA15   76   CHAIR,MEDIUM COMFORT
                                          DES15    2   DESK,1200MM (W) 600MM (D)
                                          FIR11    1   FIRE-FIGHTING EQUIPMENT,CO2 EXTINGUISHER
                                          KIK11    1   KICKSTEP
                                          MCR11    1   MICROFILM READER
                                          STA21   11   STAND,DOUBLE-SIDED,2300MM (H) WITH 200MM SHELVES
                                          STN12    1   STAND,HATS AND COATS (PARALON) SET OF 20
                                          TAB12    7   TABLE,1000MM (W) 600MM (D) WITH HEIGHT FOR STANDING WORK
                                          TRD11    2   TROLLEY,BOOKS
                                          VDU11    1   VISUAL DISPLAY UNIT


2304    OFFICE, ACQUISITIONS              ASH11    1   ASH TRAY,DESK TYPE
```

Figure 2.5 A computer-produced equipment schedule

by linking the architect and the computer by means of graphics rather than by textual and numeric communication.

Even now, graphics remain expensive because of the power of the computer needed and the cost of the ancillary equipment. Given that the capital investment can be made, however, it is now easily possible to make useful savings in time and effort in this application area. The viability of computer graphics has come to the architectural profession later than most. This is partly because it is difficult to use the drawing as input to another process. Sutherland has written, rather derisively, that: 'Pen and ink or pencil and paper have no inherent structure. They only make dirty marks on paper'[25]. This is true enough, but unfortunately the architect does not really need an inherent structure to a drawing. It is the drawing itself that contains the bulk of the information and not what can be deduced from it. This is not the case in, say, aircraft design, where a drawing of a fuselage or a wing can be used as the basis for many complex calculations on airflow and structural strength; or in manufacturing, where the drawing of a component can be used to produce a tape that will control a machine to make that component. By these means, many important savings can be made that have few direct equivalents in architecture.

The principal way the architect can derive benefit from computer-assisted graphics is therefore the improvement they give in draughting speed. Some comparative tests have been done, and the consensus is that a typical system will give a 2-3 times improvement in speed over manual drawing[26].

There are many different ways by which the computer can speed up draughting. At the simplest level, standard drawing elements, such as doors, sanitary fittings, columns, or staircases can be drawn once and then placed on the drawing as many times as necessary by specifying the position and orientation. The components can also be 'mirrored', so making a left-hand swing door into a right-hand swing.

At the next level of complexity, whole areas of a building can be repeated in a different location, or an entire floor can be repeated to form the basis of another similar floor.

The computer allows a drawing to be split into 'overlays' that can be drawn in any combination and at any scale. Thus the same structural plan can be used for all drawings of every scale that include that information. A building may typically have several such plans showing, for example, the planning grid, the structure, the services provision and dimensional information. This obviously saves repetitive drawing, but also means that if an alteration is made to the structure, only one drawing need be altered. This not only saves time but cuts down the chance of error.

The computer also makes alterations much easier. Lines or whole areas can be removed in seconds where the manual equivalent involves slow and messy erasures. The finished appearance of a computer drawing is always very clean and regular. A typical computer-produced floor plan is shown in *Figure 2.6*.

The growing use of computers has meant that in recent years there is a need to transfer drawings from one organisation to another. This can often be done even if the computers or their programs are different. Most large land surveyors now use computer techniques to collect their data in the field and to produce the basic drawing. Some take it further and work up the whole drawing on a computer-aided draughting system. This information can be transferred to the architect's computer and after deletion of information which is not required will form an immediate site plan which is guaranteed to be as accurate as any measurements made by the surveyor.

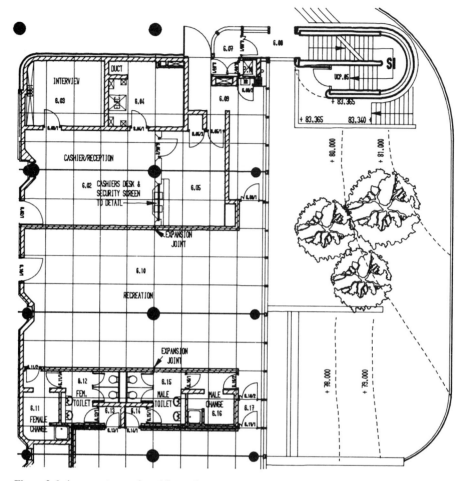

Figure 2.6 A computer-produced floor plan

In the UK, the Ordnance Survey have been transforming their maps into computer-compatible form for some years now. About 18,000 maps have been converted to date and this is increasing at the rate of 2,000 a year[27]. These maps can be bought on magnetic tape and can then be enlarged to the required scale and a site plan cut out as needed. *Figure 2.7* shows a portion of one such map.

Another aspect of computer-aided drawing is the production of perspective views of buildings. A model of the building can be input into the machine and outline views produced from any point. These views are often rather 'thin' for client presentation, but they form an excellent basis for a finished drawing. *Figure 2.8* shows a perspective of William Morris's 'Red House' produced by an advanced program that can represent textures and shadows.

Services engineering

Because of its increasingly complex nature, services engineering is not normally of direct concern to architects, although fairly recently they were responsible for this

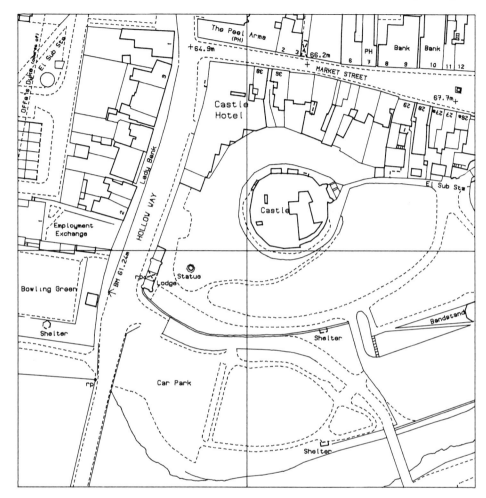

Figure 2.7 A portion of a computer-held Ordnance Survey map (Crown copyright reserved)

aspect of design. Fortunately, most of the assessments required in this area are completely objective. This is therefore one of the fields in which computer aids can be most useful. Thus although the architect cannot dispense with the services consultant, it is at least possible to check that the building will be economic to run. This will give the architect tighter control over the design and will speed the whole detail design stage, as fewer meetings with the consultant will be needed.

There is a wide variety of programs available, reflecting the range of the service engineer's work. There are, for example, numerous programs to aid the design of pipe and duct networks. In a large building, the heating, ventilation and air-conditioning ducting take up a lot of space and should ideally be integrated with the design at an early stage. At present, the architect tends to make provision for ducting and the services engineer has to modify the design later. The result is often an awkward and time-wasting compromise.

The design of drainage networks is another task that is subject to strict rules with few subjective criteria, and is therefore suited to computer solution. Programs are

Figure 2.8 A computer-produced perspective (courtesy McDonnell Douglas Information Systems)

available that will model the terrain from given spot heights and will lay out and cost a drainage network in a few minutes. When planning a housing estate, the cost of this aspect of the design is a factor that can be taken into account where previously the architect would probably have been forced to ignore it.

Environmental control is a vital consideration for the architect but is an area that demands lengthy calculation. Here again, the computer is well suited to the task. The computer can easily calculate daylight factors, design artificial lighting layouts, check glare factors and predict solar gain. Most of these matters affect the design at an early stage and should be checked before the shape of the building has hardened. *Figure 2.9* shows a typical output from such a program giving the variation in the temperature of a room with the day of the year and the time of day. From this, the architect can see where problems of heating or cooling are likely to occur.

Most important from an economic point of view, the computer can calculate the size of the heating plant necessary. Heating is now very expensive and can be the biggest factor in running a building. The computer can give the size of heating plant needed and an estimate of the running cost. Programs are available that can do this on a 'broad brush' basis, given only the general massing of the building, so being appropriate to the sketch design stage, or can do it at a very detailed level, taking into account all gains and losses through the fabric, gains from occupants or machinery, the thermal inertia of the structure etc.

Office management

One of the most valuable things that a computer can do for a large project is control it more efficiently. Because of the scale of operations, each task has to be prepared

ESPSIM: 3-D plot
ROTNXX = 45.0 ROTNYY = 45.0
YSCALE = 1.795 ZSCALE = 1.548

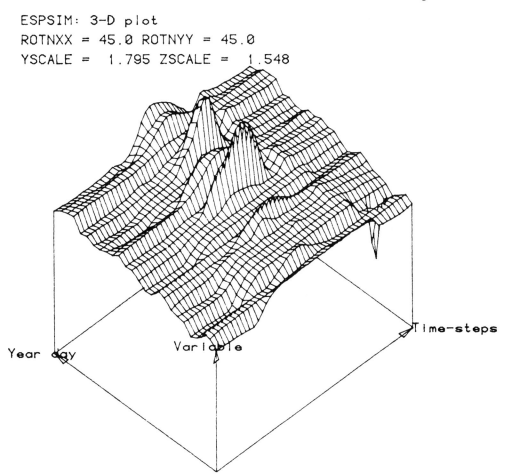

Figure 2.9 An air-temperature output (courtesy ABACUS)

for, introduced at the right time, and finished on schedule if large amounts of time and money are not to be lost. Also, a record must be maintained of the cash flow to ensure that the job is within budget. This is the sort of task that computers are well suited to, as it requires the calculation and analysis of many separate inputs. It is also the sort of task that architects feel they have too much of.

The well-known critical-path technique is widely used for controlling projects. The job is broken down into a number of separate tasks to be carried out in a certain order. By analysing the network in conjunction with the men and resources available, it is possible to predict the period within which activities should start and finish. Potential bottlenecks can be pin-pointed, accurate forecasts of how many staff will be needed at each stage can be made and the on-going cost of the design to the office can be calculated. A barchart showing the timing of activities on a certain construction project is shown in *Figure 2.10*.

An architectural practice can also check on the budgeting of the job as it affects the practice. Analysis of staff worksheets and expenses can tell the chief architect how much time and money has been spent on a job and allow early action to be

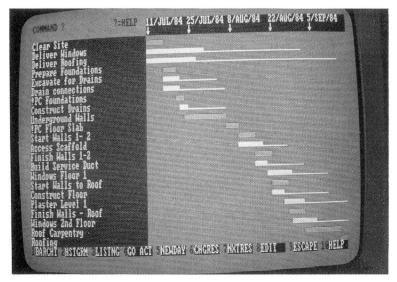

Figure 2.10 A job-programming barchart (courtesy Abtex Software Ltd)

taken. Balancing the office costs against a fee prediction can also be invaluable in forecasting the future financial position of the office. It is of course possible to do all this by hand, but is much quicker, cheaper and more accurate by computer.

Computers on a typical job

To show how these techniques might be applied in practice, it is worthwhile to trace the progress of a typical job, of say 3,000 square metres of floor area, through an office that has some familiarity with computer methods and has invested in one of the more powerful microcomputers, although not one that can produce drawings. A job of this size might reasonably require a team of four to run it for most of a design period extending over six months.

When the job is introduced, it must first be fitted into the overall office administrative structure. This is done initially by recording the existence of the job on the accounting system program. Whenever the job number is subsequently encountered in time sheets or expense accounts, the cost will be automatically totalled and matched against the predicted expenditure.

The job is also fitted into the office's work schedule by running it through a critical-path analysis program. The user will have available several 'standard' sets of data for jobs of different types and size ranges. The standard sets will become available in the normal course of events as jobs go through the office and will only require minor changes to fit them to a specific project. A work programme will be produced and will help the smooth running of the office.

Computer methods save perhaps three days in administrative tasks initially and may save up to two man-weeks of such tasks over the entire job.

When the brief is available, the records are checked to find similar jobs that have been processed by the computer. As most offices tend to specialise in a limited range of building types a reasonable match will normally be available after a year or two of computer working. The information in the archives will include room data schedules for each different room type, giving such information as the room area, its environmental requirements, special finishes or services required and a list of equipment contained. After initial editing and reprinting to suit the new job requirements the sheets will be very useful to the designer and might save a week in the initial studies period.

At the next stage of the project, the architect should determine the general approach to layout, design and construction to provide the client with outline proposals. The computer can help in this by running simulation models of parts of the design, such as circulation patterns, lift use and the demands on various other facilities. The overall shape and orientation can be tested for excessive heat loss. Cut-and-fill optimisation can be performed to help determine the most economic siting. It is unlikely that much time will be saved by using these methods but they bring the intangible benefit of a better understood design.

At the end of this stage the architect is ready to produce a scheme design. The room data schedules will have been worked up and amended by the client and the specialist consultants and can be used as a check list in preparing the sketch plans. The reduction of the need to refer to earlier jobs or to the library or the consultant will save a few more days here. During this period a copy of the room data schedules should be made for the archives, to be used as the basis for future jobs. It will be found more useful to have rather general statements on the schedules than the very specific descriptions they will acquire later.

At the next stage the detail design is prepared. Computer analysis can be used here in a more specific manner, to check such factors as window sizes, artificial lighting provision and acoustic performance.

This may actually add a few man-days to the time taken, because otherwise the designer would probably have made an estimate on the basis of experience and the detail refinement would be done by a consultant. The advantage of using the computer is that it provides more insight and more control.

The next stage is the preparation of production information and it is here that most savings can be made. Schedules of doors, windows, room finishes, sanitary fittings, etc. can be prepared with greater ease and can be rationalised with greater convenience. Savings of about five man-weeks can be hoped for at this stage.

The job is now largely handed over to the quantity surveyor or cost engineer for the preparation of bills of materials and for costing. Obviously, the improved production documentation will save the consultant a great deal of time, perhaps as much as six man-weeks. However, as the fee scales are usually fixed this does not benefit the architect and it can be argued that the architect has been doing the consultants' work for them and has also been assuming the responsibility if there should be any mistakes in the documentation. Despite this, the fact that this stage will be shortened is valuable in itself as the total design time will be reduced. Also, the increased responsibility is the inevitable concomitant of the architect's taking over more of the total design process, which is of immense importance in the long term.

It is at this point that the job is often found to be too expensive. Normally, this would mean that the architect would have to do more work to slim down the building, without being entitled to extra fees. The computer can help here by allowing the architect to find where the most lavish provision has been made, and by giving an indication of the effects of reducing it. Simulation studies could also be re-run to check for the possibilities of reducing the size of circulation spaces and other facilities. Rationalisation exercises can be carried out easily to reduce the ranges of components used or to omit their provision entirely. In all, two man-weeks could be saved and the design will suffer the least harm.

At the tender action stage, the contractors and sub-contractors will find their estimating easier and quicker as the information is more accessible than in the past. Here again, the advantages of the computer are indirect and in the form of speeding up the job rather than in saving money.

In the supervisory stages of the job, the computer will be of less use to the architect but the advantages of good documentation and a high degree of rationalisation will continue to be felt throughout the life of the project.

Overall, therefore, the savings in the example given total perhaps ten or eleven man-weeks out of a total of two man-years as summarised in *Figure 2.11*.

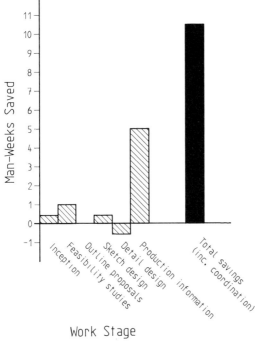

Figure 2.11 Typical time savings made by computer aids

A ten percent saving may not seem a spectacular improvement, and in fact it is true to say that the use of computers will not give savings much higher than those realised by a good rationalisation exercise or an economy drive. However, the advantages of using the computer lie not only in the saving of time but in the fact that the design will be better than could have been produced manually in a limited period. In the longer term this will presumably enhance the office's reputation and lead to more commissions.

So even a job of average size can benefit from computer aids. On a larger job, the savings will be more than proportionally greater as communication losses become significant with manual methods and co-ordination is more difficult. Greater investment in machinery can also bring benefits; using the computer to take over most of the draughting, for example, will make large savings in manpower.

Chapter 3

Equipment

Digital computer organisation

Computers come in a wide range of sizes and speeds, from small machines that can fit into a pocket, to enormous machines that need a large air-conditioned room. Despite this variation they all have the same organisation which is illustrated in simplified form as a block diagram in *Figure 3.1*.

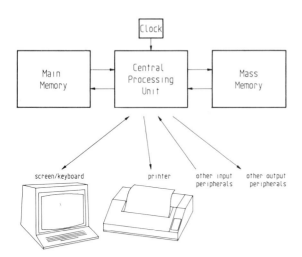

Figure 3.1 Digital computer organisation

What might be called the brain of the computer is the central processing unit (CPU). This carries out the instructions required in one elementary step at a time. At one step it might add two numbers together; at the next step compare two numbers for equality and so on. Obviously, the construction of CPUs varies widely. Some can operate much faster than others and some can handle large numbers in a single process which others must break down into smaller components.

The CPU is sensitive to the pulses coming from a timing device called the 'clock'. Each time the clock sends out a signal, the CPU carries out one instruction. Clocks are, of course, set to run at the highest speed that the system can tolerate reliably.

A small computer might have a clock running at less than one million cycles per second (usually abbreviated to megahertz or MHz) while a large machine can run at fifty times that speed or more.

The CPU can 'remember' only a few numbers at a time. Therefore, it needs to be closely connected to a memory area. In the memory resides both the program being executed and some or all of the data it needs. The CPU can take information from the memory and place information into it, possibly overwriting a previous value. In operation, the CPU will take an instruction from a location (or 'address') in memory; carry it out then go to the next location for the next instruction. The instruction could itself be the reading of a data item from a specified memory address or the writing of a data item produced by a CPU process into the memory. *Figure 3.2* shows in diagrammatic form a simple example of this process used to calculate a rectangular area.

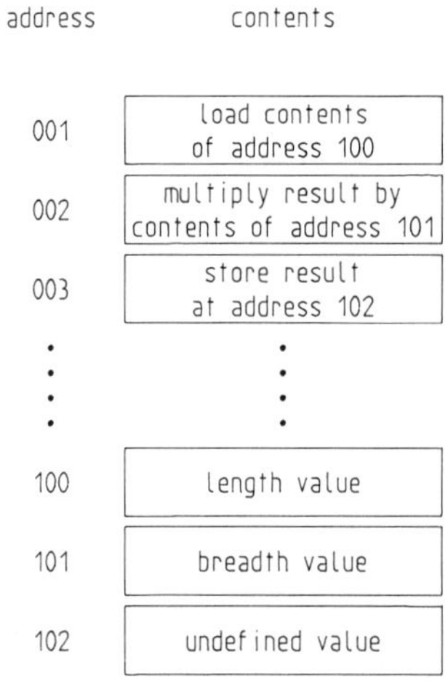

address   contents

001	load contents of address 100
002	multiply result by contents of address 101
003	store result at address 102
⋮	⋮
100	length value
101	breadth value
102	undefined value

Figure 3.2 A calculation process

The CPU is shown as initially referencing address 001. At this location is an instruction ordering the retrieval of the data item at address 100, where the length value has been stored. The CPU carries this out, then steps on to address 002. This contains an instruction ordering the multiplication of the number in the CPU with the number at address 101, which contains the breadth. The CPU then steps on to address 003 where the instruction requires the storage of the number in the CPU into address 102, overwriting the previous contents. That address now contains the area.

Note that it is not possible for the computer to distinguish data from instructions: both are held as groups of numbers. If the CPU instruction counter should be allowed to index data, as can happen if there is a fault in the program, then nonsensical results will be obtained.

An important type of instruction is the 'branch' or 'goto' command, which tells the CPU to take the next result from a specified memory location, rather than the next in series as it would normally do. This facility is very useful because the CPU can be made to go back to an earlier part of the program and execute it over again. The ability to 'loop' in this manner means that a single piece of program can process any amount of data.

Alternatively, the branch instructions can be used to choose one program path rather than another. This will be done on the basis of the result of a previous process. So for example, an accounting program might total the expenses on a job, then subtract that figure from the fee prediction. If the result were negative, indicating a loss on the job, a branch could be taken to a separate section of the program that would print a warning message.

The combined use of looping, branching and modification of data in memory gives the digital computer a very high degree of flexibility.

The memory described above can be read from or written to at any random point. It is therefore called Random Access Memory or RAM. Most modern computers also have certain amount of Read Only Memory or ROM. This type of memory contains programs or values which are permanently fixed and cannot be altered by the CPU. Such programs are typically used for carrying out standard functions which are frequently required.

Memory size is measured in characters, or 'bytes', one byte being, for example, a single letter of the alphabet. For reason of convenience to the computer, memories are built in multiples of 1024 bytes, this being rather confusingly called a kilobyte, or Kb for short. A larger unit is 1024 kilobytes which is called a megabyte, or Mb.

The size of memories varies widely. A small microcomputer might have only 16 kilobytes of memory, while a very large commercial machine could have hundreds of megabytes.

The built-in memory of a computer suffers from several drawbacks. The first is that being built in, it cannot be removed and put into an archive. Random Access Memory is also normally 'volatile', that is, all the information is lost when the machine is switched off. Further, even a memory of millions of characters is often insufficient in itself: it could only encompass a medium-sized database, for instance. For these reasons, computers are fitted with 'mass memory' which can take a number of forms.

The cheapest and slowest type of mass memory is magnetic tape. This works on exactly the same principles as are used in a domestic tape recorder. A serious limitation of tape storage is that it is 'serial access' rather than 'random access'; i.e. to obtain a certain piece of information it is necessary to read through all the information previous to it on the tape.

For efficient regular use, various types of magnetic disc are the most popular form of mass memory. Essentially, all such devices consist of one or more discs coated with a magnetic medium which are spun at high speeds. The reading and writing 'heads' are moved in and out to access bands of information. It is therefore a random access device. This system is illustrated in *Figure 3.3*.

Discs can store a great deal of information, tens or even hundreds of megabytes, and can transfer it at high speeds, although this is still only a fraction of the speed of main memory access. Traditionally, the CPU and main memory are considered the essential core of the computer system and all other devices are referred to as 'peripheral devices' or just 'peripherals'. The mass memory system is therefore a peripheral although these days it is often contained within the same casing as the CPU and main memory. The other important peripherals are those that the user needs to send data to the computer and receive answers back. When using microcomputers, the most common input peripheral will be the keyboard and the most common output peripherals a screen and a printer. Data transfer on all these devices is much slower than that of mass memory transfers and operates in a rather different way. Early machines used the CPU to control peripherals, but modern

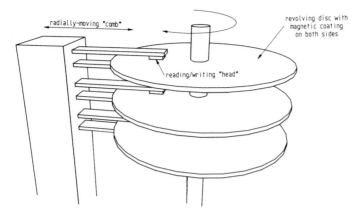

radially-moving "comb"

revolving disc with
magnetic coating
on both sides

reading/writing "head"

Figure 3.3 Disc organisation

computers have special interface circuits which can carry out a certain amount of simple processing in order to remove the load on the CPU. The circuit assigned to the disc memory, called the 'disc controller', is usually a very complex system in itself.

Mainframes, minicomputers and microcomputers

The distinction between mainframes (very large commercial machines), mini-computers and microcomputers is to a large extent notional. All digital computers work broadly in the way described above, the differences being principally those of scale.

A mainframe computer is often very large indeed. Typically, it will be contained in a number of separate cabinets each several metres long. The CPU will be a very complex affair which understands hundreds of different instructions and executes them at high speeds. Computer speed is commonly measured in millions of instructions per second (mips) using standard 'mixes' of equivalent elementary instructions for comparison between machines. The fastest IBM mainframe at the time of writing is the 3090/200 which operates at 25 to 27 mips[28] and mainframes are available that will run at over 200 mips. Such machines of course cost far more than any architectural practice could contemplate. *Figure 3.4* shows a typical mainframe installation.

When minicomputers arrived in the early 1970s they were intended to be relatively slow and simple devices for controlling specific applications. However, the increase in their speed and complexity over the years has meant they are now considered relatively powerful machines. A typical minicomputer will be contained within a single cabinet of one or two cubic metres in size. Its speed will be about one-tenth of that of a mainframe computer. The fastest machine currently made by Prime Inc., whose computers are widely used in the computer-aided design field, is the model 9955 which can execute 4 mips[29]. A more typical speed for a minicomputer would be 0.5-1.0 mips. The extra power that minicomputers provide can be essential for large-scale applications and many architectural practices have invested in one. They are however, relatively expensive. A computer system including peripherals will cost about the equivalent of ten man-years in salaries.

Figure 3.4 A mainframe installation (courtesy SIA Ltd)

Figure 3.5 A typical minicomputer installation (courtesy Prime Computer (UK) Ltd)

Figure 3.5 shows a typical minicomputer installation with separate cabinets containing a disc drive and tape drive.

The most important characteristic of a microcomputer is that its CPU is contained in a single integrated circuit, or 'chip', which is called a microprocessor. The first microprocessor was built by the Intel Corporation in 1971 and was initially used to control devices such as lifts or slot machines. Intel and other firms began producing more complex versions and by the late 1970s the first microcomputers were built by teaming a microprocessor with other standard chips.

A modern microcomputer is typically about the size of a typewriter and incorporates within a single case a keyboard, a printer or screen and a disc drive. Its speed can vary from about 0.1 mips on machines with a slow microprocessor to over 1 mips on a 'supermicro' such as the Universe 68/05 which is manufactured by Charles River Data Systems[30].

Microcomputers are not only restricted in speed, but also in the amount of main memory that they can support. Even some expensive microcomputers cannot support more than 256 kilobytes and some have a maximum as low as 16 kilobytes. This restricts the complexity of the programs used.

The cost of these machines is very low by commercial standards; a relatively fast 'business' microcomputer will still cost only the equivalent of one or two months' salary. *Figure 3.6* shows a popular business microcomputer.

Analogue computers

The computers described so far have been of the type known as digital computers, because they hold and manipulate information in the form of exact numbers, and this type is by far the most flexible and the most widespread. There are, however, several other principles on which to build computers, the best known of which is the analogue principle. An analogue computer uses different amounts of some physical quantity to represent numbers; thus a slide rule could be considered as a simple analogue computer because length represents a value. In practice, an analogue computer will almost always use electrical voltage for its representative medium.

Analogue computers are programmed by plugging leads into a patchboard to link different pieces of circuitry together. They are faster than digital computers because the problem is solved as a whole rather than one step at a time. However, their accuracy is limited and they can only be used for certain very restricted classes of problem, notably those involving differential equations.

Digital computers are much more versatile than analogue computers, and therefore suitable for a greater variety of applications.

Computer bureaux

A few years ago, the great majority of architectural practices that used a computer did so through a commercial bureau. The bureaux owned and operated very large mainframe computers and customers could use the machines either by calling in person with their problems in the form of a deck of punched cards or a magnetic tape or remotely by using a teleprinter terminal and a telephone link.

Since computers have became more affordable, only a small number of practices continue to use bureaux. However, bureaux can offer some advantages. Very

Figure 3.6 A typical business microcomputer (courtesy IBM (UK) Ltd)

complex analyses such as those often encountered in structural engineering could take many hours to process on a microcomputer, where a large mainframe would need only minutes. In this case, a large machine is a necessity if the design team is not be delayed. It is also an advantage that the customer is only charged for the time that is used, so a fluctuating workload may be more economically handled at a bureau. The problems of maintenance and obsolescence are borne by the bureaux, so relieving the customer of many minor worries, and the staff at a bureau can offer high expertise in their fields[31].

Despite this, the advantages of having a personal computer under direct control outweigh the disadvantages for most architectural firms, where the problems are many and varied but not particularly large in scale.

Mass storage peripherals

The two important types of mass storage, tape storage and disc storage, can appear in a number of different forms. Tape and disc are similar in that they both work on the principle of moving a magnetic medium past a reading or writing head, but as has been pointed out, the more complex organisation of disc storage allows it to access data more quickly.

Although tape is much slower than disc, its very low cost and high capacity has meant that it is still very popular for archival storage. The tape used by minicomputers and mainframes is almost always a half inch wide and commonly comes on reels in lengths of 600, 1,200 or 2,400 feet. Depending on the recording mode, a 1,200 foot tape can store about 50 megabytes.

A very important advantage of reel-to-reel tape is that standard sizes and recording modes are universally in force. So a tape can be written on one computer and taken to any other that has a tape drive. This is not possible with disc storage. There is a wide variety of physical shapes of disc, and even for a given shape, the way the information is written will be different between machines. Transfers are only possible within a given manufacturer's range, and often only for certain models within the range.

The tape used with personal microcomputers is usually in cassette form. The capacity of a cassette is limited, but is usually at least 16 kilobytes. Many microcomputers can use a domestic cassette recorder as the tape drive, thus providing a very cheap storage system. The disadvantages are that cassettes are not normally interchangeable between computer systems because of the different ways in which information is written, and, more important, that they are extremely slow. It can take five minutes to load the contents of a cassette into main memory, and while this sort of delay may be acceptable when using computers as a hobby, it is not practicable in a business environment.

Disc drives can be categorised in several ways, but one division is into those that use flexible, or 'floppy', discs and those that use 'hard' discs, which have a rigid substrate. Floppy discs were originally introduced by IBM and had a diameter of 8 inches, which is still a very popular size. *Figure 3.7* shows a floppy disc.

Later developments brought in the 5¼ inch, the 3½ inch and the 3 inch diameter versions, all of which are currently in use[32].

A floppy disc has many advantages. It is extremely cheap. It has a high storage capacity: an 8 inch disc can hold about 1 megabyte. It is very compact, and it is very rugged. Its disadvantage is that it is only moderately fast; a good drive will transfer data at about 64 kilobytes/second, but this is quite adequate for most microcomputers, where the limitation will probably be the CPU speed.

'Hard' discs come in many configurations. They may consist of a single disc, or many discs sharing a common spindle. They may be removable or permanently sealed within the computer. They may be of widely different diameters and run at very different speeds. However, for minicomputer or microcomputer use, two

Figure 3.7 A floppy disc

different types predominate. These are the 'cartridge' disc and the 'Winchester' disc.

Cartridge discs consist of single disc sealed inside a plastic container of about 400 mm diameter. The cartridge is inserted into the disc drive in different ways according to its type and can be removed for archived storage.

Winchester discs are discs permanently sealed in an airtight environment. This has the great advantage of excluding dust particles.

Hard disc systems are relatively expensive but have the advantages of speed and high capacity, both of which are about ten times that of a floppy disc. A single disc can store from 10-30 megabytes and the drives can transfer data at up to a megabyte/second. They are therefore useful for applications which require the handling of large amounts of information, especially database management.

A problem with such high speed devices is that they are less reliable than slower systems such as floppy discs. At its perimeter, a hard disc can be travelling at 40 metres/second and the reading and writing heads are suspended just 0.003 mm above it. This gives very little margin for error: as the diameter of a human hair, for example, is 0.08 mm it is not uncommon for a catastrophic failure to take place in which the head makes physical contact with the disc and severely damages it. This 'head crash' is an expensive business. The disc will be a write-off and the drive itself will need extensive repair. Worse, the information on that disc will be lost. *Figure 3.8* illustrates the fine tolerances involved.

Whichever form of disc system is used, it is important to have a second mass storage system so that copies can be made for archival storage. Typical arrangements are two floppy discs; a Winchester disc and a floppy disc or a Winchester disc and a tape drive.

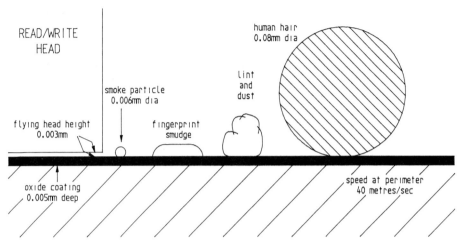

Figure 3.8 Disc tolerances

Input peripherals

The most basic input device, and one that is available on all computers, is the keyboard. The layout is very similar to that of an ordinary typewriter keyboard although certain symbols, such as fractions, will not be present. The user can type instructions to the computer or enter data as required. The computer almost always accepts information on a line-by-line basis, so the user can type a line, correcting any errors by using the backspace or 'rubout' key, and it will not be processed until the carriage return key is pressed.

Many keyboards have a separate numeric pad containing the ten digits and the decimal point to facilitate purely numeric input. It is also usual for keyboards to be fitted with special 'function keys', the number of which varies. These keys can be set up to be equivalent to a whole string of separate characters. So for example it would be possible to set up a function key to be equivalent to the word 'architect', after which a single press of that key would transmit the whole word. It can be very useful to store commonly used commands in this way. *Figure 3.9* shows a typical computer keyboard with a numeric pad on the right and a row of function keys along the top.

It is frequently necessary to indicate positions on a screen and pass them to the computer. This might be necessary in a draughting program to indicate the end points of a line for example, or in a word processing program to indicate a character to be deleted. It is possible to do this via the keyboard by assigning four keys to move an indicator on screen up or down and left or right; repeated keystrokes can then be used to index any character position. This solution is easy and cheap and is in fact commonly used in word processing and other applications where a display of alphanumeric data has to be edited. It suffers from the drawback that it is slow, especially where high resolution is required, as in graphics work. A whole range of devices have therefore been developed to carry out this task.

The most direct positional input devices are those where the user actually touches the screen at the point required. Some screens are made 'touch sensitive' by a grid of fine wires laid across their surface; the placing of the user's finger on a

Figure 3.9 A typical keyboard layout

wire alters its electrical characteristics and the position can therefore be detected. This method is too crude for many applications, however, as it impossible to define a point with any accuracy.

Another direct positional input device is the 'lightpen' which is a pen-like device that has a light-sensitive tip. Most modern screens constantly redraw the picture many times a second, so that when the user touches the screen the time that the pen detects a light is the time when that part of the picture is being drawn. The lightpen is not a particularly popular device these days, partly because it is still not very precise and partly because it is uncomfortable to work for long periods holding the pen to the screen.

The most popular positional input devices are those that work indirectly. The screen displays a large cross (or 'cursor') on its surface which shows the point currently being addressed. The user can then move this cursor by using a device that is held in the hand. By feeding back observations of the position of the cursor into movements of the hand it is easy to find any point on the screen. It is then necessary to press a separate button to tell the computer that the final position has been reached.

Perhaps the most common indirect positional input device is the joystick. This consists of an arm that moves freely standing on a fixed base. When the user moves the arm from the vertical in any direction, the cursor also moves in that direction. A joystick may or may not be 'progressive'; that is, greater deflections from the vertical cause the cursor to move faster. A progressive device is faster in use but needs more practice. Joysticks often incorporate a 'firing button' which is used to confirm the final position of the cursor. Some joysticks can also be rotated, which makes it possible to define a 'third dimension'. Joysticks are widely used in the hobby section of the market, but are rarely found in commercial use these days. This is because getting to a very precise position requires a series of minute 'nudges' of the joystick and it takes some time to acquire this skill.

Devices that are similar in principle to the joystick are the 'trackball' and 'thumbwheels'. A trackball is a ball fitted into a fixed base; as the user spins the ball in any direction, the cursor also moves. A trackball is illustrated in *Figure 3.10*.

Figure 3.10 A trackball (courtesy Gresham Lion Ltd)

Thumbwheels are a pair of small wheels set at right angles to each other with a portion of their rims uppermost. One wheel controls the left to right motion of the cursor and one the up and down motion. By using the thumb and first finger of one hand the user can reach any position on the screen. *Figure 3.11* shows a pair of thumbwheels set into a keyboard.

Figure 3.11 A pair of thumbwheels

Joysticks, trackballs and thumbwheels operate on a fixed base, but there are also indirect positional input devices that operate freely. One such device that has become very popular for use with microcomputers is the 'mouse'. This is a small hand-held unit that is fitted with two wheels set at right angles to each other. As the user moves the mouse over the desk in front of him, the wheels spin and send signals back to the computer. The mouse has one or more buttons set into it to confirm positions. *Figure 3.12* shows a mouse.

Figure 3.12 A mouse (courtesy Spicers Ltd)

Another free device is the 'tablet' or 'digitiser'. With this device a special pen containing an electronic coil is moved across a pad in which is embedded a grid of fine wires. The wires detect the impulses from the pen and pass back its position. A tablet is much more expensive than other forms of positional input but has several important advantages. One advantage is that it can be used to trace over existing maps, diagrams or drawings. This is not possible with a mouse due to wheel slippage. It is also very accurate: a typical tablet can return a point to the nearest

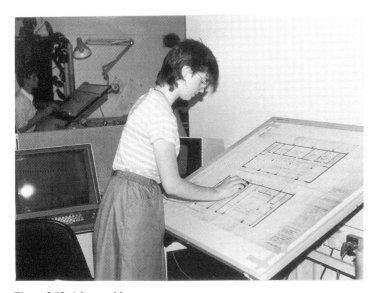

Figure 3.13 A large tablet

0.2 mm, which is at least as high as the accuracy of the guiding hand. In order to take greater advantage of this precision it is possible to replace the pen by a 'puck' which is heavier and incorporates a magnifying lens. Tablets come in various sizes from A4 versions to A0 or larger. *Figure 3.13* shows a large tablet in use.

A further advantage of a tablet is that it can have a dual function as a text input device. This is done by the use of a 'menu'. In computer usage, a menu is a list of displayed alternatives. By touching a certain alternative, an instruction or item of data associated with it will be accepted by the computer. *Figure 3.14* shows this system in use. A small tablet has a card glued to its surface which contains many alternative commands including at the top a full alphabet and set of digits. The tablet is fitted with a puck which in this case has four confirm buttons. Pressing one button invokes the menu command under the puck, while pressing another sends back the current position of the cursor on screen.

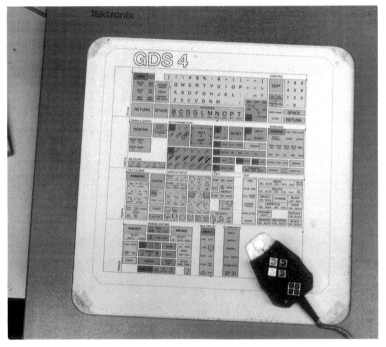

Figure 3.14 A small tablet and menu

It is also possible to have a menu displayed on screen and use any positional input device to select from it. This has the disadvantage that some of the screen will then be unusable for other purposes, and this is a serious drawback with a small screen. It does however, have the advantage that the menu can be 'dynamic'; that is, selecting a certain alternative will clear the current menu and replace it by a menu of further choices now open to the user. For example, when using a draughting program the user might touch the alternative marked 'draw'; a menu would then be substituted that offered a choice from 'line', 'rectangle', 'arc' and 'circle'; if 'circle' were chosen, a menu of the ways in which a circle can be defined would be displayed, including 'radius and centre', 'three perimeter points', 'radius and tangent' and so on. This solution can guide a user through a complex program with the minimum of confusion and is widely used.

A type of menu which has found favour recently is the 'icon' menu. In this, instead of the alternatives being written out, small pictures are offered. Thus instead of the word 'draw', a picture of a pen could be displayed; instead of the words 'line', 'rectangle', 'arc' and 'circle' the appropriate diagrams could be shown[33]. This is a more 'friendly' system and is international. It does however, use more screen space than a typed menu and certain concepts cannot be convincingly represented by a picture.

Although there is a no 'best' positional input device, the lightpen, mouse and tablet are faster in use than the fixed-base units. The lightpen is not ergonomically satisfactory and the tablet has advantages that other devices do not, but is much more expensive.

The keyboard and a positional input device are probably all that most users will require. However, many different data capture peripherals are available, some of which can be very useful in the right circumstances. One of these is 'optical character recognition' (OCR) input[34]. An OCR machine scans printed text and sends it to the computer automatically. At one time, only special typefaces could be read, which made the devices of very limited use, because if text had to be prepared in the special typeface it might as well be input directly into the computer. However, modern machines can read most typefaces in common use. Such a machine can be very useful when an architect has to input text from other sources. For example, collating data from a catalogue to store in a database, or when converting a manual system into computer form. Unfortunately, OCR machines are still relatively expensive.

The conversion of existing graphic data into computer form presents similar problems. When using a computer-aided draughting system it is often necessary to input existing drawings which were produced before the computer was introduced or which come from other sources. Machines exist to scan such drawings and turn them into numerical coordinates which are sent to the computer. The results are never quite as good as the originals, because of the limits of scanning resolution and of the quality of the original drawing, and normally need some editing once on the computer. Most such machines are very expensive indeed, but in our own office we have developed and marketed a low-cost scanner which uses a plotter to move a light-sensitive head over a drawing[35]. As a plotter is necessary to draughting in any case, it provides the precision transport mechanism at no extra cost. *Figure 3.15* shows the unit in use.

Output peripherals

The most common output peripheral is the printer. All installations, even those that work mainly through screen output, will need printed output from time to time. There is a number of different principles on which printers can operate[36]. The very fastest types work on electrostatic principles and print an entire page at a time at a speed of up to 21,000 lines per minute. Also very fast and much more widespread, are 'line printers'. These machines have a complete set of characters for each print position along a line. The line is set up, then printed in a single impact, hence the name. Line printers work at about 1,500 lines per minute.

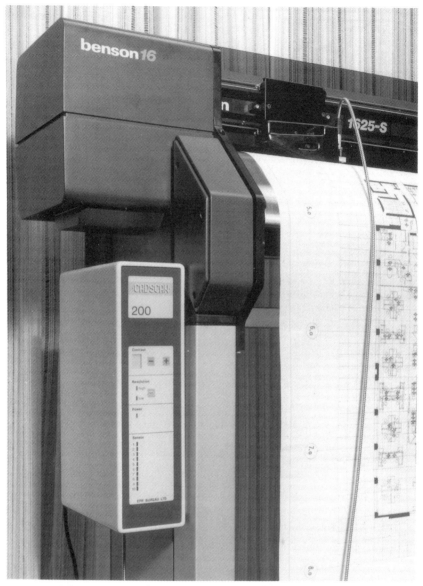

Figure 3.15 A low-cost scanner (courtesy EPR Bureau Ltd)

Such printers are only appropriate to mainframe computers and to large-scale data-processing. Architectural applications will not normally require a lot of printed output and slower and cheaper printers will be adequate. Low-speed printers have a fixed paper carriage and a moving head that prints one character at a time, in the same way as a modern typewriter. The most common speeds available are 30 characters per second and 120 characters per second. The higher speed is the more useful as it can take up to three minutes to print a single page at 30 characters per second.

A popular type of printer is the dot-matrix printer. This machine has a printing head which contains a number of needles arranged typically as nine rows each of seven pins. Different sets of pins are pushed forward to strike dots through an inked ribbon and form a character. These printers are fast and relatively cheap. The big drawback with the basic versions is that the output is not of correspondence quality. More complex and expensive dot-matrix printers are available which can give better quality output by overprinting each character several times, moving the print head slightly each time. This does of course slow the printing speed correspondingly.

The other popular form of printer, which produces correspondence-quality output, is the daisywheel printer. This uses a printing head which has all the characters arranged around a central core, rather like the petals on the daisy. To print a character, the wheel is spun and the uppermost character struck on to the paper through a ribbon. *Figure 3.16* shows a daisywheel.

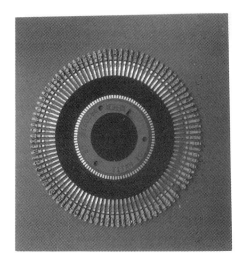

Figure 3.16 A daisywheel

These devices are slower and more expensive than dot-matrix printers, and also more noisy. However if only one printer can be justified it is often the best choice because of the importance of word-processing in any business.

A type of printer that is often provided with personal microcomputers is the thermal printer. This uses heat-sensitive paper and a row of needles across the width of the paper. Heating selected needles will form an image out of dots. This device is very cheap and almost totally silent, but because of its low speed and the cost of the special paper it is not appropriate to business use.

Among the less common printers are the 'ink jet' printers. These fire a continuous stream of minute droplets at the paper. The droplets are ionised so that magnetic fields can be used to direct them to the correct positions. Ink jet printers give good quality output and are also capable of producing graphics as well as text. If several jets are present, these graphics can be in colour. One of their main disadvantages is the tendency of the jets to clog up if the printer is not used for any length of time, and this can make them too unreliable for some installations.

Another type of printer which has been introduced only recently is the laser printer. The principle of this device is that a moving mirror is used to direct a laser

---

Architectural Design	Issue #12	Winter

Arch News

**The Society of Architects
12 Golden Square
London W2**

Value is what the 1985 SOA National Seminar in London is about — and it's what you'll find in greater quantity and greater quality than ever before at the convention's 1985 SOA Exhibit of New Technology and Products.

The biggest convention exhibit in SOA history will present you with a demanding challenge: Find a way to decide which products, services and technologies you want to learn more about.

- Computer Technology
- Software, Hardware
- Contract Furniture
- Interiors, Exteriors
- Practice Systems

You'll find it all — plus a convention schedule that lets you see it all — at the 1985 SOA Display of New Products and Design.

The Art of Dormers: A retrospective look at 19th Century dormers at the Birmingham Architectural Gallery, through January 12.

*"Architecture...
has it's own
validity. It needs
no reference to
any other disipline
to make it viable
or to justify
its value."*
— Philip Johnson

Master Architect

The man featured in this month' issue may well be one of Oxford': best kept secrets. You may no know his face, but if you live ir Oxford you know his work — tha is, if you've ever visited civic anc residential buildings. The man i: Arthur Erickson, Architect, and h(has called Oxford home for mos of his life.

While the layperson may no recognise his face or name, durin; a remarkable and prolific caree spanning more than 30 years Arthur Erickson has receivec dozens of honorary degrees anc virtually every major professiona and personal award. To list then all would take pages, but the) include the Man of the Yea award 1972 and the Tau Sigm; Gold Medal for excellence ir design.

Conferences, seminars, workshops:

May 21-25: 25th Annual Meeting The Society of Architects, 12 Golden Square, London. W2.
June 10: Deadline, call for 500-word abstracts, "Designing and Managing Commercial Buildings: An Intensive Workshop."
June 25: British International Solar Conference and Exhibition. Thomas Convention Center, London. NW1.

Call 01 962 7171 for details and register early for the 1985 SOA National Seminar, March 9-12, in London.

*"Architecture
aims at eternity;
and therefore is
the only thing
incapable of
modes of fashions
in its principles."*
— Christopher Wren

This page was created with Aldus PageMaker and the Apple LaserWriter printer.

continued

Figure 3.17 Output from a laser printer (courtesy Apple Computer (UK) Ltd)

beam onto a drum with a special surface and so generate points of magnetic charge. These points attract particles of magnetised ink which are then transferred onto the paper. Laser printers are about two to three times the cost of a good impact printer, but are fast and silent and are capable of producing graphics as well as text. *Figure 3.17* shows an output from a popular laser printer.

The other output peripheral in common use is the screen, or 'visual display unit' (VDU) as it is often called. The type of screen most used works on the 'raster scan' principle, as does a domestic television. The image is formed out of many dots (or 'pixels') which are continuously being redrawn, usually at a rate of fifty times per second. Personal microcomputers normally have a special piece of circuitry driving a socket which has an output that mimics a television signal. The information about the dots is taken from a section of main memory. These computers can therefore be plugged into an ordinary television set with a consequent saving of money. Screens built specifically for use with computers contain the memory to record the image within themselves and access it to refresh the picture displayed. They can therefore use a much simpler and slower link to the computer.

When used in its simplest fashion, the screen will be displaying alphanumeric text. Usually 24 lines of 80 characters can be shown at one time.

At a more advanced level, a screen will be required to display graphics. With a raster screen this is not conceptually very difficult. Each line has to be broken down into dots and the image memory altered. Again, this is done by the computer if it has a television link, or by the screen itself if it is a more advanced model with that capability. The displayed image, however, immediately highlights the drawback of a raster screen. Because the image is formed out of dots which have fixed positions, lines drawn at an angle will have a 'stepped' effect. The closer the line is to the horizontal or vertical, the more apparent this effect becomes. *Figure 3.18* shows how this effect occurs.

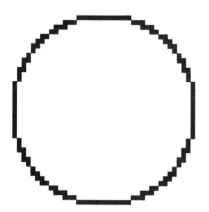

Figure 3.18 Stepping effects on a raster screen

Because each dot on the screen has to have a corresponding record in memory, it becomes expensive to build a screen of high resolution. A typical cheap screen, for instance, might have a resolution of 600 dots by 250, so 150,000 memory locations will be needed. This resolution is only suitable for crude graphics such as bar charts or graphs, however. To be able to show a reasonable amount of an architectural drawing in a readable form cannot be done with much less than 800 by 400 dots, which will require well over 300,000 memory locations. Not only does the memory become expensive, but as conventional television technology cannot be used, the cost savings of mass production are lost.

A line drawing will look crude to the eye at any resolution below 1024 by 800 dots. The best raster screens available at present can go up to 2048 by 1586 dots, but the price of these high-resolution units is typically the equivalent of one or two man-years of salaries and this could not be justified by most practices. *Figure 3.19* shows a powerful and popular, but relatively expensive, screen with a resolution of 1024 by 768 dots.

Figure 3.19 A Tektronix 4111 raster screen (courtesy Tektronix UK Ltd)

The raster screen is now easily the most popular type of screen because of its inherent advantages and because the cost of the large amounts of memory it needs has become so much less than before. There are however, other types of screen which have their own advantages and are still in use. The most famous of these is the storage screen, which was developed by Tektronix Inc. This was once the 'workhorse' of computer-aided draughting, but is now largely obsolescent. With this system, the image is drawn on to a fine wire mesh behind the screen which is then used to activate the screen phosphor. As the wire mesh itself keeps a record of the picture information, this type of screen is relatively cheap to build even for high resolution. Lines do not have to be broken down into dots, so there are no stepping problems and very fine images can be obtained.

The big disadvantage of this type of screen is that it is impossible to erase individual lines. The only way to remove a line is to clear the entire screen by sending a charge through the wire mesh and redraw the picture with the unwanted line absent. This can be very slow and irritating to the user. A less important restriction for architectural use is that the image is monochrome, while raster screens can support colour.

Another type of screen which has become less popular in recent years is the 'vector refresh' screen. With this system, the screen contains a memory, but instead of recording the image as dots, it retains them in the form of the original lines making up the drawing. What is in effect a simple computer inside the screen continuously reads through this 'display list' in memory, interprets it and draws the lines on screen. The advantage of this system is that, like the storage screen, very high resolution is not prohibitively expensive. Selective erasing can be made simply by removing the information from the display list. Also, the more advanced screens can manipulate the picture without placing any load on the computer. For example, the picture can be scaled up or down, or selected pieces of it can be displayed. This can be done in seconds whereas reference back to the computer might involve several minutes of delay.

The cost of this type of screen is less than a raster screen of equivalent resolution but the use of a display list does impose a restriction. This is that the complexity of the picture supported may not be unlimited as it can be with a storage or raster screen. Eventually, the number of lines in the picture will exceed the size of the display list and some of the picture will not be drawn. However, the best of the modern vector refresh screens have a great deal of memory. I have found by experiment that a very complex floor plan can be contained within 128 kilobytes of memory, and modern vector refresh screens often have many times that amount.

Certain of the most advanced raster screens also incorporate display lists. The list is not continuously scanned as with vector refresh systems, but only as needed to update the raster memory which indexes the individual dots. This gives the advantages of quick scaling or manipulation of parts of the image without placing a load on the computer.

When working with graphics it is eventually necessary to produce the drawing on paper so that it can be issued. The fastest and most direct way of outputting a drawing is to record the image on screen directly. This is done by a 'hard copy unit' which can take the information sent to the screen and use it to drive a mechanical device. Various systems are in use, but a popular one uses heat-sensitive paper. *Figure 3.20* shows such a unit.

A hard copy unit will produce a drawing in seconds, but it will not be of very high quality and more importantly only A4, or sometimes A3, copies can be output.

For full-size drawings a plotter is required. The most popular type is called a drum plotter. The paper runs over a cylinder that can rotate in either direction. The pens are positioned above the drum and can move along its axis in either direction. Thus by combining the movements of the drum and the pen lines can be drawn in any direction. Drum plotters are relatively cheap, but an A0 size will still cost the equivalent of a year's salary. *Figure 3.21* shows a popular drum plotter.

The other widespread type of mechanical plotter is the flatbed plotter. With this type, the paper is static and the pen alone moves. Flatbed plotters are more accurate, although the accuracy of most drum plotters is quite sufficient for

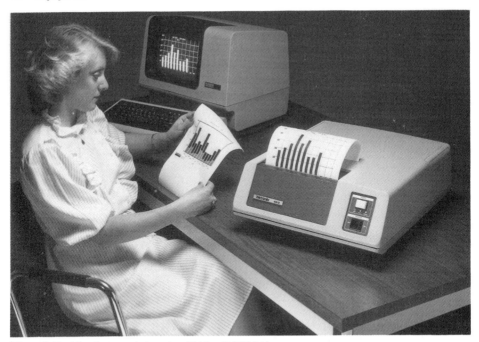

Figure 3.20 A hard copy unit (courtesy Tektronix UK Ltd

Figure 3.21 A drum plotter (courtesy Benson Electronics)

ordinary purposes, and can use any type of paper. The larger sizes are however usually much more expensive than drum plotters and take up a lot of floor space.

For mechanical reasons most plotters of both types are limited to movement in one of eight or occasionally sixteen directions. Lines which do not correspond to any of these directions are approximated by combinations of the standard direction in steps of typically 0.05 mm in length. Such a step size will give a very acceptable result, but cheaper plotters will use a larger step size and then lines which are very close but not quite concurrent with a standard direction will be visibly jagged.

Plotters are surprisingly primitive in use and it will be found that a disproportionate amount of time is spent in making sure they run smoothly; in fact a number of installations have taken on a junior employee who does nothing else. Paper movement is one major problem. The plotter may for instance draw a grid line, then perhaps half an hour later come back and draw a wall line that lies along the grid, but because of paper movement appears a millimetre or two away from it. Precise control of temperature and humidity or the use of rather expensive plastic film will counter this.

Ink pens give the best line, but are very susceptible to drying up. If this is not noticed and corrected in seconds the entire plot may be useless. Different coloured ball-point or felt-tip pens are available which rarely dry up, but because lines can only be of a single thickness correct photocopies or dyeline prints cannot be made.

Plotting a large drawing is a slow business. Even a moderately complex floor plan can take several hours, so it is important to plan ahead when organising drawing issues. It is also important to check that the computer can be used for working on a drawing or for other purposes at the same time as it is driving the plotter. Some of the more primitive machines cannot do this, so when using them either a lot of time is lost or plotting must be done overnight, which is not particularly satisfactory.

Many of the problems associated with plotters can be overcome by the use of an

Figure 3.22 Electrostatic plotters (courtesy Benson Electronics)

electrostatic plotter. These machines use many thousand of needles laid along the width of the paper. Each needle is able to control a dot of ink. The best of the modern electrostatic plotters have over 500 needles to the inch and this gives a result that is virtually indistinguishable from a pen plotter. *Figure 3.22* shows one such device.

These machines can produce a complete A0 drawing in three or four minutes, so logistical problems almost disappear and producing extra plots for checking and marking up becomes more practicable. They also have the advantage that they are quiet in use, whereas a pen plotter needs a sound-proofed room. They are also capable, given the right software, of producing half-tone pictures. *Figure 3.23* shows an interesting example.

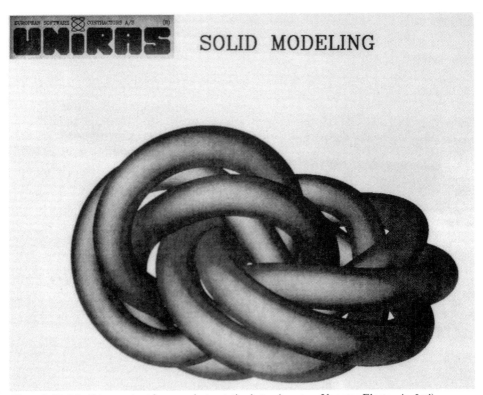

Figure 3.23 A half-tone output from an electrostatic plotter (courtesy Versatec Electronics Ltd)

Some electrostatic plotters incorporate several ink baths and so are capable of producing full colour output at up to A0 size.

The biggest disadvantage of electrostatic plotters is their sheer cost. They are several times the cost of pen plotters and justifying the amount spent demands a heavy workload and tight deadlines to be met.

Communications

It is often assumed by inexperienced users that there are few problems in passing information between computers. In fact, this is one of the most difficult areas. It has already been pointed out that it is not possible simply to take a disc from one disc drive and put it in another, unless the drives and the associated computers are identical. Reel-to-reel tape is freely interchangeable between computers, but not many microcomputers are fitted with tape drives.

The usual method of communicating is by connecting together standard sockets (usually called 'ports') on the two machines. These ports are made to international standards which lay down the physical dimensions and the way in which information is to be passed. The standard which is by far the most widely used is the RS232C interface[37]. This uses a D-shaped plug and socket with 25 pins. *Figure 3.24* shows an RS232C interface.

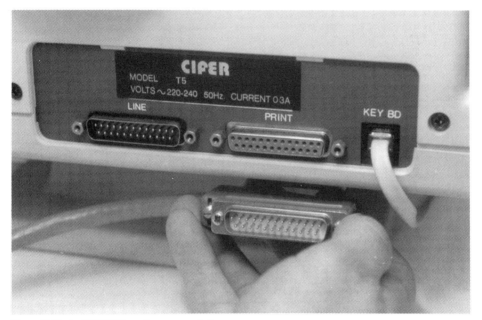

Figure 3.24 An RS232C interface port and plug

Other interface standards are the IEEE488, which is used on a few microcomputers, and the Centronics interface which is often used by printers to receive data. It is possible to buy adaptors that make communication between different interfaces possible.

The ports are also the usual methods by which the computer sends and receives information from the peripheral devices. A small microcomputer may have only one port and so only be able to communicate with a single peripheral, usually a printer. A larger machine might have up to, say, 16 ports.

Data may be sent at various speeds and the connected devices must be set to the same rates. The interface standards lay down the possible rates, of which the most common are 300, 1200 and 9600 bits per second (usually called baud). In most cases, it takes ten bits to transmit a single character, so 300 baud is 30 characters per second. Even 9600 baud is slow compared with the speed at which a computer can generate data and is only suitable for use with the slower peripherals, or for relatively small amount of information or for occasional transfers.

The simplest way of connecting two ports is of course by a fixed cable. This will be reliable up to at least 100 metres, and 'line drivers' are available that will extend this up to thousands of metres. A typical minicomputer installation will have a central computer and cables running from it to screens scattered about the office.

Obviously, communicating with devices outside the building cannot be done so directly. In these cases, the public telephone network can be used. A device called a modem (modulator/demodulator) is needed to accomplish this. The modem plugs into an interface port, takes the data and converts it into the audible sounds which travel down the telephone wire. It performs the reverse process to receive data. The faster modems are wired into the telephone, but a more convenient, if less reliable, type for many purposes is the acoustic coupled modem. This device has two cups into which the telephone handset is placed. This means that it can be used with a portable computer to send back data from site or the architect's home. *Figure 3.25* shows an acoustic coupled modem.

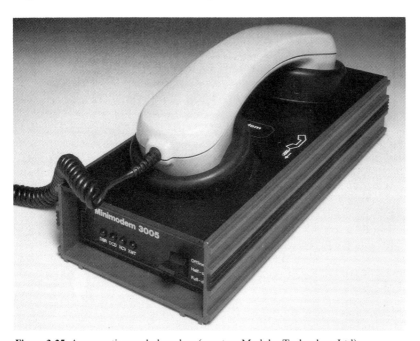

Figure 3.25 An acoustic coupled modem (courtesy Modular Technology Ltd)

The main problem with modems is that the public telephone network is not reliable for data transmission. A low speed of 300 baud is usually used and although some work at 1200 baud or above, they need to incorporate systems for automatic error checking and retransmission when necessary.

Linking through interface ports is not suitable when large amounts of data have to be transferred. This is often the case when some equipment is shared. For example an office might have several relatively cheap microcomputers, but only one plotter or correspondence-quality printer. Another situation is where a single disc is shared by several computers; this is necessary when they all need to access a common database, as might happen for instance when several designers are working on different aspects of a single job.

In the simplest case, when up to three or four computers need occasional use of a single expensive peripheral, a multi-way switch which is set manually is a cheap and effective solution. In more complex environment, a local area network (LAN) will be necessary[38]. A local area network connects a group of computers with high speed data links. A typical microcomputer network will work at between 62,000 characters per second and 1,200,000 characters per second, which is hundreds or thousands of times faster than using an interface port. Typically, up to 32 computers can be linked within a single network. In use, the fact that data goes through a shared network is invisible to the user, who will merely send or request information in the usual fashion.

A local area network consists physically of a special piece of circuitry added to each computer that will transmit and receive data, and the cabling that links the machines together. The program that controls the computer is also modified so that it recognises that the machine is part of a network. The network may be cabled in various ways, the more popular of which are diagrammatically illustrated in *Figure 3.26*.

The 'star' system is conceptually the simplest, where a single master machine controls the shared peripherals and services the subsidiary machines as required. The 'ring' system has all the computers on a closed loop and data is passed around the ring. An 'address' at the front of each set of data ensures that it is ignored by all machines except the one chosen to be its destination. The 'bus' system has all the computers connected to a central spine which carries the data. It works in much the same way as a ring system, but is more reliable as the failure of one machine will not necessarily stop the whole network.

Local area networks can also be classified as using 'baseband' or 'broadband' transmission methods. Baseband networks carry data on a single channel while broadband networks can use several channels at once and so are faster. Various proprietary local area network packages are available for all popular business microcomputers.

It should be noted that networks are only necessary when several independent computers are in operation. An office might well decide that a simpler solution than purchasing a network is to buy a more powerful single machine which can service a number of users through its interface ports.

Choosing a computer

The newcomer to the field will be baffled by the number of computers available; there are hundreds to choose from, all with their own advantages and drawbacks.

It will be found on close inspection that the choice is more narrow than it might appear at first sight. This is because there are only a few makes of microprocessors available and manufacturers choose one or the other and add their own cabinets and features to make up the complete computer[39]. They are however, constrained by the capabilities of the microprocessor, so the final machines cannot be very different in function and will in general accept the same programs, occasionally after some minor modifications.

At the slower end of the market, the Zilog Corporation Z80 and Z80A microprocessors are very popular. At the time of writing, over one third of all microcomputers use one of these chips. They are however, rather slow for many business purposes. The firm of MOS technology make a range of microprocessors

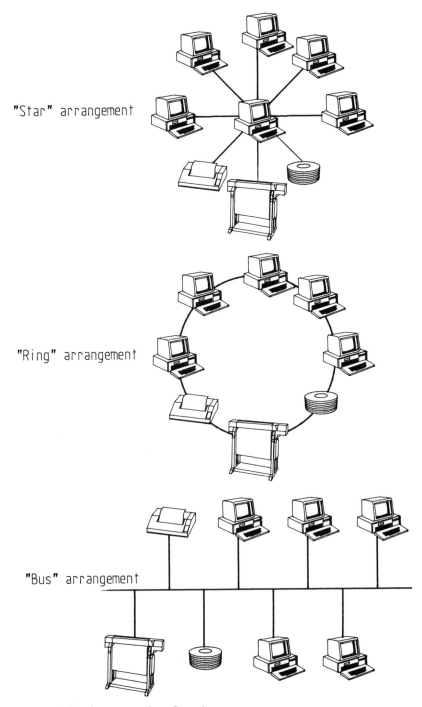

"Star" arrangement

"Ring" arrangement

"Bus" arrangement

Figure 3.26 Local area network configurations

of which the most popular are the 6502 and the 6510. Again, these are mainly used by the slower machines.

In the middle range, the Intel Corporation is dominant; the 8088 processor is used on many machines, including the IBM PC. The 8086 processor is also very popular. The fastest 'supermicro' market is dominated at present by the Motorola Corporation with their 68000 and 68020 processors.

The power of a microcomputer is shown partly by its clock speed. A microprocessor that is clocked at less than about 4 megahertz will be found rather slow for business use. The same microprocessor may well be clocked at different rates in different computers: this depends on the quality of the components which communicate with the processor. By using more expensive ancillary chips a manufacturer may be able to squeeze more speed out of a computer.

The other indication of microprocessor power is the size of numbers which it can deal with. Most of the slower machines have '8-bit' processors. This means that only numbers containing up to 8 binary digits ('bits') can be handled. A bit can be the number zero or one, so the largest number that can be expressed is 2 multiplied by itself 8 times; that is, 256. This does not mean that the computer itself cannot handle numbers larger than this; larger numbers are broken down into smaller chunks and the result obtained by carrying out several calculations rather than one. This process will be invisible to the user, but obviously overall speed will be reduced.

Business microcomputers usually use 16-bit microprocessors such as the Intel 8088. The use of 16 bits allows numbers up to 65536 (or plus or minus 32686 when negative numbers are used) and in practice this size is sufficient to run most processes efficiently. Most minicomputers also have central processing units which operate on a 16-bit basis.

The very latest and fastest microprocessors such as the Motorola 68020 are 32-bit devices, which is as large a size as that used by the majority of mainframe computers. This size will deal in a single operation with numbers that have fractional parts. Thirty-two bits can be used to express as wide a range of numbers as will reasonably be needed to a precision of six significant figures.

The number of bits a microprocessor can handle is important for another reason; it controls the amount of main memory that can be accessed directly. Most microcomputers use 16 bits to address memory and so up to 65536 locations (64 kilobytes) can be supported directly. Most 16-bit business microcomputers can however run programs larger than this. They accomplish this by dividing the total memory into 64 kilobyte chunks and swopping between the chunks as necessary. There is a obviously a certain overhead in doing this. In practice, 64 kilobytes is enough for most of the programs that will be run on microcomputers, if not for the most complex, some of which use a megabyte or more.

Eight-bit processors use two numbers to make up 16-bit addresses, so can access as much memory but take an extra clock cycle to do so. In contrast, the 32-bit microprocessors are capable of directly addressing far more memory than would ever be fitted to the computer.

A different solution is provided by the Intel 80286 which is used on the IBM PC/AT. This uses 16 bits for its internal arithmetic, but 24 bits for addressing memory[40].

It is wise to choose a computer based on a popular microprocessor, as then a wider range of software is available. A study of program lists will help to narrow the field as programs will only run on certain processors. Having decided on the microprocessor, the individual manufacturers can be considered. Each will offer various facilities of which the most important is the disc drive. It is very desirable that the disc is of high capacity; a megabyte is a reasonable minimum and ten megabytes is comfortable for most microcomputer applications. The disc speed is also important, especially when database handling is being done as there will be a lot of transfers between the disc and the main memory. It is more important here to compare 'access times' for the discs, which is the time they take to find a piece of information, rather than the 'transfer rate' which is the rate at which the information will be passed after being located.

In certain applications, a 'lap held' microcomputer will be found to be the best choice. A lap held computer is a lightweight machine that is powered by internal batteries and so can be taken anywhere. This can be most useful for site visits or working at home. Because it is essential that they be compact and fully transportable, lap held computers must make compromises elsewhere. For example, instead of a screen, an LCD display of only four or five lines may be provided. The only form of mass storage likely to be available is a miniature cassette tape of the type used in dictating machines. These are slow and of limited capacity. On the other hand, lap held machines often have relatively sophisticated communication facilities. This means that they can be used to gather raw data in the field, then to transfer it to a larger computer for further processing. A good example of this application is in measuring up a building. The bearings and distances can be typed in and the computer will make basic checks for consistency. Later, the data can be passed to a larger machine that will generate a drawing on a plotter. *Figure 3.27* shows a popular lap-held microcomputer.

When considering a minicomputer rather than a microcomputer, the problem of choosing a machine will not arise in the same way. The usual situation will be that

Figure 3.27 A lap-held computer (courtesy Epson (UK) Ltd)

an office will decide to buy a program, typically a draughting system, which requires a larger machine. As minicomputers all have different central processing units, the program will only be available on one, or perhaps two, different makes of computer. In the computer-aided design field the firms of Prime Inc. and the Digital Equipment Corporation are very popular.

The various machines in a manufacturer's range will be of different power, but usually a program written for one will run on all machines in the range above the one it is designed for. The problem then becomes which of these to choose. The smaller machines will only be able to support one or two users without the response time becoming poor, while the larger, more expensive versions might support up to a dozen. The user can only make an estimate of the amount of use that is likely to be made of the machine, trying both not to underestimate capacity nor to pay for excess, especially bearing in mind that at the present rate of change a computer will become obsolete in three or four years at the most.

Office installation

At one time, all computers needed to be kept in sound-proofed rooms, because the noise of their cooling systems was so high. They also needed air-conditioning, because they dissipated so much heat, and their power consumption was so high that a three-phase supply was necessary. Modern microcomputers are not so demanding. They are very quiet; the heat generated by a computer and screen is about 125 watts, little more than a light bulb, and they can be plugged into an ordinary power socket. Even quite powerful minicomputers can now be fitted into an office environment. A typical mid-range minicomputer will have a noise level of 60-70 DbA and dissipate 1000-2000 watts[41].

Noise is still a problem with mechanical peripherals. Printers, especially high-speed printers, have a noise level that is barely acceptable. It is however, possible to buy acoustic hoods to cut this down. Fast pen plotters are too noisy to be acceptable to most offices and will need their own room for this reason and also because they require close temperature and humidity control to prevent paper movement.

The quality of the electrical power supply is a problem. The ordinary supply is not 'clean' and will be subject to spikes and drops in voltage caused by other machinery on the same cable switching on or off. To fit clean supply points especially for computer use causes disruption and is rather expensive and inflexible. Unfortunately, it is almost a necessity for reliable working. Without a clean supply the computer can be expected to mysteriously 'crash' regularly. This is the term used for it becoming disorientated and need to be reset and restarted. A crash will normally be only a nuisance and only lose at worst a few hours of time in running a session. Occasionally, however, it will happen at a critical time when the computer is updating a disc and then all the information on that disc will be lost. When this happens, the user has no option but to go back to the last archive copy (if there is one) and recreate perhaps days of work.

Small and cheap power filters can be bought which will remove spikes and surges of current. These will help, but will of course not cure the problem of a drop in voltage. Complete power failure is always a potential hazard when an office uses computers; fortunately it is a relatively rare occurrence. It is possible to buy uninterruptable power systems which use banks of batteries that take over when the mains fail, but these are very expensive both to buy and to service and will not normally be contemplated by an architectural practice.

The use of anti-static carpeting is a wise precaution. This will remove the risk of static charges disorienting the computer, is not expensive and looks well. This type of carpet also has a bonded pile that will not dislodge and cause a potential hazard to discs. Floppy discs are more tolerant to dust particles than other types, but a reduction in air contamination is still worthwhile.

The comfort of the user must always be borne in mind as computer sessions can require mental concentration for hours at a time. The chair should be of the fully-adjustable type used by typists. There should be adequate layout space for drawing and documents and glare should be avoided. Modern raster screens are reasonably bright, so it is not too difficult to avoid glare from the source documents being illuminated at a much higher level than the screen. Reflections from windows or light fittings can be very tiring on the eyes and care should be taken to angle the screen to prevent this. Detailed studies of the ergonomic and medical problems of visual display units have been published by various bodies and can be consulted for detailed information[42].

Chapter 4

Programs

Availability

No matter how powerful the computer, it is useless without programs to control it, just as a car is useless without a driver. However, it is not always easy to obtain useful and reliable programs. After 20 or 30 years of development, this might seem a surprising statement. It might be expected that by now most of the basic programs would have been written and become well-known as their use spread. The problem is that the machinery is developing so fast that the software is being left behind. Computers are continually becoming faster and cheaper, so a program that was written, say, three years ago that took full economic advantage of the existing machinery would now seem unacceptably primitive and lacking in advanced facilities. More elaborate peripherals are also always becoming available. The introduction of cheap colour screens, for instance, has had a marked impact on programs. Very many of them will now use diagrams to present results where they would previously have printed columns of numbers.

In the more general business field, the situation is not too bad. As the potential market is so large, a good deal of money and effort can be justified in developing software. Popular and effective programs can be found fairly easily in the fields of accounting, database management and word processing. In the more specifically architectural field, it is much more difficult to find programs. The profession is not a rich one nor a particularly large one. Also, the range of problems an architect deals with is very wide, covering as it does the arts, the sciences, the crafts and the integration of all these to produce the finished building. Consequently the range of programs needed is also wide. Progress is always being made, but the number of programs available is still not particularly large.

The problem is made worse by the lack of compatibility between computers. A program written for one machine will not normally run directly on another. A difficult conversion process will be necessary and a continuing investment will be necessary to keep all the versions up to date. For this reason, a user who has purchased one of the less popular computers may find the choice is further limited.

Origination

New users often imagine that they will write many or most of their own programs to solve their particular problems. In principle this is a good idea because a program can be tailored to the user's exact requirements and to the limitations of a particular installation. In practice, the user rapidly discovers that writing a program of any size is a very slow process. A relatively simple program of 1,000 instructions would take a good programmer several months to write and correct. As the size of the program goes up, the time needed goes up disproportionately because of the interactions within the program. Most advanced programs these days incorporate many man-years of development and considerations of cost and lead time usually rule out purpose-written solutions for the average office.

Organisations exist that will write programs to order, but despite the gains of professionalism they are extremely expensive. Most users will find that they regularly write small programs to get over day-to-day difficulties, and that they will often swop programs with other architects. Most of their work will be done using proprietary software, however.

Ready-made programs, usually called 'packages', are now in widespread use. This was not always the case. A few years ago a good package would cost the equivalent of several man-years of salaries. Fortunately, with the vast increase in computer ownership that has come with the microcomputer, the number of program copies sold has increased and the price has fallen to such an extent that it is now common practice for an office to purchase programs on a 'throw-away' basis. Several different programs will be bought which carry out the same task; they will be tested and the less suitable ones discarded.

The penalty paid for the use of packages is that the user must change from the former manner of working to the standard method laid down by the software designer, and this may conflict with the normal office procedure. It may also be so general, in order to be able to handle a wide range of cases, that it is inefficient or clumsy when dealing with a more restricted set of problems. A good package, however, will be usable in a natural sort of fashion that sometimes turns out to be better than the procedure the user has adopted up to that time.

Portability

Programs are usually listed according to the computer they run on. This can sometimes be misleading because a very important factor in the microcomputer field is the type of microprocessor that the computer is built round. As was pointed out in the last chapter, although there are many different makes of microcomputers they are based on a relatively small number of microprocessors. Each type of microprocessor has a different set of basic instructions. It is therefore necessary to find programs that will run on a particular type of micropocessor.

In some circumstances it may be possible to adapt a program to run on a different machine. This is made feasible by the fact that most programs are written in a 'high level' language. High-level languages have internationally agreed forms and each manufacturer provides a type of program called a 'compiler' or an 'interpreter' that converts a program written in that language into 'machine code' instructions that can be executed by a particular processor.

By far the most popular language in the microcomputer field is BASIC[43]. The name is an acronym of Beginner's All-Purpose Symbolic Instruction Code and, as the name suggests, was originally developed as a training aid. It became very popular, however, and developed in sophistication and is now used for quite large programs. BASIC is not suitable for the most complex programs, as it does not allow itself to be broken down into self-contained modules.

For more advanced programs a language called FORTRAN (Formula Translation Language) is often used[44]. Most graphics and analysis programs use this language. A popular language for artificial intelligence work is LISP (List Processing Language) which is used by many expert systems[45].

There are other languages available, usually biased towards a particular type of problem. The difficulty is that no two compilers work in exactly the same way. All manufacturers have their own dialect of a particular language and detail changes have to be made throughout when converting a program. More seriously, some parts of a program will often be written directly in machine code, or the 'assembly language' which translates directly into machine code. This is done to take advantage of facilities which cannot be accessed through a high-level language or to increase the speed of the program.

The other factor that hinders the transportability of a program is the 'operating system' it runs under. All computers are equipped with a controlling program that oversees the function of the machine and schedules the tasks to be carried out[46]. It is also the direct interface to the user and accepts his or her commands. Programs need to communicate with the operating system and so a particular type of operating system must be present for one computer's programs to be compatible with other computers.

Many manufacturers issue their own operating systems, but there are some generally accepted systems. The very popular Zilog Z80 and Z80A microprocessors often run under an operating system called CP/M (Control Program/ Microcomputer). More recently, a system called MSX has been adopted as a standard by some manufacturers but has had a mixed response. The IBM microcomputers use a system called MS-DOS and this has been taken up by many computers which are compatible with IBM. Among the most advanced machines, a system called UNIX is finding favour[47]. UNIX may not be easy to use initially, but is very powerful in its capabilities when throughly understood.

The other factor that can prevent a program from being used is the facilities it needs. A program will need a minimum amount of memory, for instance, and this may be above the basic amount supplied by the manufacturer. It may also require certain peripherals such as a disc drive, screen and so on. Obviously, the user must consider if it is worth while purchasing additional features.

A few popular programs are available for more than one make of computer or microprocessor, but in general software suppliers find the effort of keeping up with a single manufacturer's developments is costly enough. The more advanced and complex the program, the more this will be the case.

Information sources

Sources of information are now widely available. The computer manufacturers issue catalogues of programs suitable for their machines and these not only include the standard business utilities but often are broken down by subject area, including those designed for architects.

Guides to programs specific to the construction industry are also available[48], and there are now even guides that list and comment on programs dealing with specific subject areas such as daylight analysis[124] or critical path techniques[98]. It is unfortunately inevitable that by the time guides have been collated and published they are well out of date.

User groups and societies have existed for years and many more have sprung up in the wake of the mass use of microcomputers. In the UK the Construction Industry Computing Association (CICA) is an old-established organisation and in the USA the Society for Computer Applications in Engineering, Planning and Architecture fulfills a similar role. There are many smaller groups that exist to share experiences and programs between users of particular systems. Membership of these can be invaluable in gaining information on particular hardware or software. Independent advice can also be obtained from consultancy firms and organisations[49].

Exhibitions and conferences are held each year with the emphasis on particular subjects including computer-aided design, computers in architecture and computer-aided graphics. The very latest developments will be presented at such conferences, but this may often mean that the programs described are still experimental and not suitable for direct use in practice.

Lists of programs of potential interest can therefore be complied quite quickly. It will be found on closer inquiry however, that many programs cannot be used in practice. The problem can be illustrated by a survey carried out some years ago to evaluate programs that produce building perspectives[85]. Of the original 30 programs mentioned in the literature, only three were actually usable. The rest were obsolete programs with no support provided, programs of purely academic interest or programs without a local agent.

Most software suppliers will be glad to send details of their products. This can vary from a single photocopied sheet of typing to a glossy full-colour brochure full of pictures of smiling users and pretty machine operators. Because this literature has to make a quick impression on users in a wide range of disciplines it cannot go into much detail, but will give a general idea of the scope of the program. The suppliers will also usually be prepared to give the names of existing users that can be contacted for independent opinions. Such opinions can be very helpful, although it must always be remembered that people have a natural reluctance to admit to having made mistakes, and the greater the investment, the greater the reluctance.

It is also important to check that the results have actually been based on practice. Very often enthusiasts fiddle with a program to try to make it useful, but it may be unsuitable for practical use.

Demonstrations can usually be arranged, but they are not always as revealing as might be expected. They will of course be organised so as to bring out the best points and gloss over the worst. Another problem is that without experience of a particular application area, it is difficult to know which features are most important and what questions to ask.

The basic data used for a demonstration will already be prepared and will be in small quantities, so it is difficult to get an idea of the effort needed to collect the data. The speed at which the program runs may be quite misleading. For example, a database program may appear to be adequately fast on a sample of data-capture, but when presented with the data on a full-size building might take hours to retrieve information, or might be unable to contain the data at all.

The demonstration data will probably also avoid awkward cases. For example a draughting program may avoid showing much text on a drawing because drawing characters is a slow process; a perspective program may avoid curved surfaces because they have to be broken down into unattractive flat surfaces and so on.

Some demonstrations go as far as to be pre-recorded onto a cassette and are played back a bit at a time with the demonstrator explaining the significance of each portion. This sort of presentation is not really acceptable: it is too easy to cut out or edit the awkward or lengthy pauses and leave an entirely false impression. Naturally, a film can give an even more misleading impression of the ease of use of a program, although a well-made film can use diagrams and close-ups to give a better idea of the scope and capabilities of a program than could be communicated at a conventional demonstration.

One failing of some vendors of computer programs is to 'sell ahead'; that is, to claim that certain improvements are being made to the program that will make it much more powerful or more useful. Such claims are probably genuine, but should be ignored. It takes a long time to write, test and document a modification properly and by then some other program may be on the market with the same capabilities. As the whole field is one of continuous and rapid development the next program will always be a better one, so a decision should be made on the basis of what is currently available.

An important point to consider with most programs is the upgrade capability. A user will typically start in a small way, but when a program proves itself it will be natural to extend its use. If this is to be possible, faster versions of the computer must be available, or possibly facilities for linking several computers together. Better and faster peripherals might be desirable to give better quality output or more convenience in use.

Considerations in use

The most important consideration in using a program is the maintenance provided. Once an office has committed itself to using a program in practice it is vital that it works and if it does not, that there is an expert to correct the fault. The purchaser usually pays a maintenance fee of about 6% of the cost of the program each year in order to secure this help. All programs contain 'bugs' which may not show themselves for a long time, until a particular combination of data will trigger the bug and prevent or corrupt the results. It is therefore necessary to check that an experienced programmer will be available at short notice. This can be a particular difficulty when buying a program from a university or technical college. Such programs can be extremely good because extensive research facilities plus a bias towards a logical approach will usually ensure a solid basis for the program. Also, as they do not have to justify themselves commercially, the programs can be good value for money. The drawback is that university holidays occupy five months of the year, and the student turnover is high, so maintenance is more often than not a serious problem.

It might be thought that an outside firm could undertake correction and maintenance of a program. This is not usually possible, because of the legal situation. The laws of copyright are unclear, but it is accepted that in most countries they have no application to computer programs[50]. This situation is changing; there have been some test cases and one programmer has managed to take out a patent in the USA (US Patent 4,270,182) covering a program[51]. A law has also recently been passed by the British Parliament that extends copyright to computer programs[52]. However, for practical reasons, all software suppliers keep their coding secret at present and will probably continue to do so in the future. This is ensured by their only issuing the program in its compiled form, when it has been translated from the high-level language into machine code and is not intelligible to humans.

The office staff will need training in the operation of the program and reference manuals will also be required so that obscure points can be looked up. Some, but not all, software suppliers will provide the training, usually at the customer's site, and a certain amount of training may be included in the purchase price. Reference manuals are often a weak point; they are not considered a priority and are often scrappy or unclear. A facility that is sometimes provided, especially for complex programs, is a telephone enquiry service. An expert user will be available for consultation and can explain difficult points or advise on the best way of approaching a problem.

A potentially very serious matter is the reliability of the results produced by the computer. If a program gives the wrong answers, there may be expensive consequences and it is then not clear who is responsible. The software supplier often requires that the purchaser signs a contract which indemnifies the supplier from any such responsibility. A case in point is the experience of a local authority in the UK when a program made errors in the design of a multi-storey car park[53]. Reinforcing had to be added later at a rather high cost. As more reliance is placed on computers, such cases can be expected to happen more frequently. The only solution would appear to be to run rough checks on any output that could be in question.

Chapter 5
Databases

Principles

The handling of large amounts of information was an early application of computers and over the years many programs were developed to deal with different categories of data. It came to be realised, however, that a single set of techniques can be used to handle almost any type of information, and by the early 1970s generalised database management systems were in common use[54].

A database is essentially a mass of information organised so as to permit easy extraction of the items of information contained. Computers do not have to be involved: an ordinary filing system is a database, as is a dictionary, although the latter cannot be updated.

To get a clearer idea of the concept of organising data, we can consider a list of clients such as would typically be used for secretarial purposes. The list might be divided into four columns, giving for each entry the client's name, address, telephone number and the name of a contact person. This list can be expressed as a tree structure having the form illustrated diagrammatically in *Figure 5.1*.

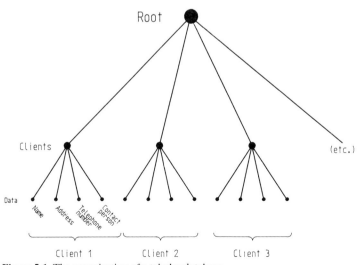

Figure 5.1 The organisation of a tabular database

From a single 'root node' there is a branch to a node for each client, from which in turn there are branches to the actual items of data. This list is a rigid structure in that there are exactly four items of data for each entry and so it lends itself to being laid out in tabular form. Many of the databases an architect will use have this form; thus a room-finishes database might have entries under columns for the room number, floor finish, wall finish and ceiling finish. Similarly, a database holding information on doors would have entries for door height, width, colour, etc.

The columns are referred to in database terminology as 'fields'. Each field will have an identification name or code by which it will be referenced and will be filled with a certain type of data. So in a door database, the field identified as 'height' will be filled with numeric data, while the field identified as 'colour' will require textual data. Various different data types will be allowed. Most systems allow a distinction in numeric data between whole numbers, such as would be needed to record the number of employees in a firm, and numbers that can have fractional parts, such as might be needed to record a dimension. Many of the better systems also allow dates to be entered directly. This can be useful, for instance in recording vital construction dates so as to aid resource planning.

The group of fields making up a single entry is known as a 'record'. Thus in the original database described, each client is a single record. All the records together will make up a single logically distinct set of information which is kept on the computer's mass storage system as a 'file'. In the simple case, the file is the database itself. Database systems will impose a limit on the number of records that can be contained in a single file and a limit on the number of fields to a single record. These limits are often restrictive; some programs allow less than 30 fields per record and under 1,000 records per file. These limitations should be checked before using the program.

Although a simple tabular structure is adequate for many databases, it cannot deal directly with the case when there are several data items associated with a single field of a record. For example, if the client database were extended so as to include a list of the jobs carried out for each client, a field would have to be added that would contain between one and, say, a dozen job names. Some database systems will allow this to be done directly and the arrangement is shown diagrammatically in *Figure 5.2*.

This hierarchical structure can be generalised to any number of levels. In the client database, for example, each job could in turn have a list of drawings issued and the date of issue. In this case, there would be another level which contained matched pairs of fields, one for the drawing number or name and one for the date.

Hierarchical structures can be difficult to handle because of their complexity. For this reason a more popular way of dealing with fields which can have multiple values is to put these values into a separate tabular database and link them together via a common field. Thus in the extended client database, a separate file could be created that contains records consisting of two fields: one for the client name and one for a single job done for that client. The client name field is used by the program to relate the two databases together and this organisation is known as a relational database[55]. *Figure 5.3* shows the arrangement diagrammatically.

Again this can be generalised to any complexity by adding more files. Another file containing records consisting of a job name, a single drawing number and a date would extend the database further, with in this case the job name field as the link. Relational databases are less elegant in use than hierarchical databases, but are conceptually much easier to handle. There are again limitations on the number of

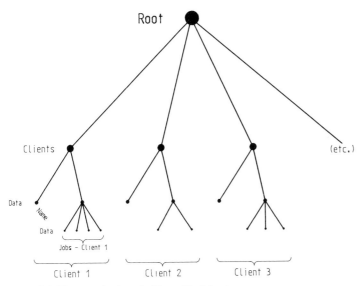

Figure 5.2 The organisation of a hierarchical database

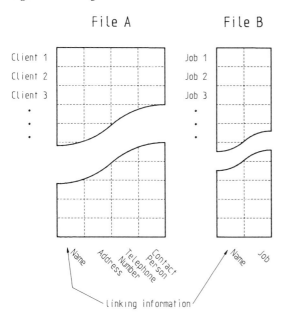

Figure 5.3 The organisation of a relational database

files that can be handled simultaneously. Some systems can only deal with a pair of files at once, while the more sophisticated can deal with dozens or more. Less than three simultaneous files can be restrictive, so again it is worth checking this point before using the program to set up a complicated database.

At the present time, relational databases are by far the most popular type of database. This is probably because they are powerful and flexible and yet easy to understand.

A facility possessed by some database systems is that they are multi-user systems; that is they will allow several users to access the same database at the same time. Such a facility is obviously essential in such fields as aircraft booking, when many agents may be checking flight availability or recording seat sales. The facility is less useful in architectural applications, but a large office might find it convenient to have a number of terminals all of which have constant free access to any project. In order to prevent contradictory situations arising the program will typically place restrictions on access at a record level. If one user is changing a record, another user may not be allowed access, or may be allowed to look at the record but not perform changes.

Operations

A database system must be capable of supporting several classes of operation. The most basic is the actual recording of the information. All systems allow new records to be created and will prompt the user for data values to fill each field. At least some checking will be performed, for instance it will not be permissible to supply alphanumeric text if the field is expecting a number. Some systems can be instructed to go further and to reject or to query values outside certain limits. So for example a door database might query door heights below 2,000 mm or above 2,200 mm. This will reduce errors due to typing.

A useful feature often available when recording data is to be able to repeat fields. Thus if the program can be told that a new door has the same characteristics as an existing one, a great deal of typing can be eliminated. It may also be possible to duplicate information in other ways. For example, it may be possible to group a number of records, then set values that will apply to certain fields in all those records. Again using the door database example, it might be possible to initialise all similar leaf types very quickly, then initialise all door fittings, such as locks and handles, in a few separate operations.

The effect of database facilities is that if the information contains a reasonable amount of repetition on any basis, it should be quicker to create a schedule by computer methods than by manual ones. In some cases it may even be possible to use an archived job as the basis for a new one to save even more effort.

The next class of operation that must be adequately supported is the output of results. Most often, a full listing of all information contained will be required. As the job progresses, the database will undergo successive updates. Without computer methods, this would mean that a schedule would become more and more illegible as it collected crossings out and amendments or would require a lot of work to retype or rewrite. With the computer, regular revisions can be issued that will always be clear and of a professional quality. Automatic titles, page numbers, running totals, cover sheets and so on are easy to arrange.

The layout of the printed information can be varied as required and this can be used to give outputs that are more readable. At the simplest level, certain fields can be omitted if irrelevant. For instance, when printing a room finishes schedule, the materials and finishes for the walls, floors and ceilings might be printed separately and sent to the various specialist sub-contractors. This shortened form with all non-essential information left out would be easier to use than the full form.

In certain circumstances, a printout of a few records or even of a single record will be required. This would allow the extraction, say, of a group of rooms or one particular room. Most systems will allow a selection of records to be printed and such a printout can form a document for discussion or issue.

One useful form of output that is not possible with manual methods is a reordered form, that is with the records arranged in the order of any field. For example, a typical schedule of equipment would consist of a section for each room and within each section there would be an alphabetical list of the equipment contained. This database could be printed in the order of the equipment description. In this form each item of equipment would have its own section, within which there would be an ordered list of the rooms that contain that item. *Figure 5.4* shows a page from such a reordered schedule, and corresponds to the schedule illustrated in *Figure 2.5*.

```
EAST CHEAM UNIVERSITY LIBRARY - FURNITURE AND EQUIPMENT
30 JULY 1986


                                           ROOM
CODE    DESCRIPTION                        NUMBER QTY  ROOM NAME


DES11   DESK,TYPIST                         2104B   1  INTER-LIBRARY LOANS


DES13   DESK,1000MM (W) 750MM (D)           2114    3  CLERICAL SERVICES
                                            2309A   1  CATALOGUE BIBLIOGRAPHY REFERENCE


DES14   DESK,1400MM (W) 750MM (D)           2101B   2  CIRCULATION
                                            2102    1  OFFICE
                                            2103    1  OFFICE,SUPERINTENDANT
                                            2104B   2  INTER-LIBRARY LOADS
                                            2114    3  CLERICAL SERVICES
                                            2202    2  BINDERY PREPARATION
                                            2203A   6  GOVERNMENT PUBLICATIONS
                                            2203B  13  CATALOGUING
                                            2203C   6  ACQUISITIONS
                                            2209    1  OFFICE, CATALOGUING
                                            2210    1  OFFICE, GOV. PUBS.
                                            3109    3  SPECIAL PROJECTS
                                            3112    3  SPECIAL PROJECTS
                                            3120    2  OFFICE
                                            3121    1  OFFICE, SECRETARY
                                            4106    1  OFFICE, KEEPER


DES15   DESK,1200MM (W) 600MM (D)           2101B   4  CIRCULATION
                                            2104A   2  ENQUIRIES
                                            2203C   2  ACQUISITIONS
                                            2301    2  GENERAL COLLECTION
                                            3103    1  OFFICE, ADMINISTRATION OFFICER
                                            3301    4  READER/STACK ZONE
                                            3305    2  STUDY ROOM
                                            3306    2  STUDY ROOM
                                            3307    2  STUDY ROOM
                                            3308    2  STUDY ROOM


DIS13   DISPENSER,SOAP                      1111    5  LAVATORY, MALE
                                            1115    3  LAVATORY, FEMALE
                                            2117    1  LAVATORY, MALE
```

Figure 5.4 A page from a reordered equipment schedule

Most sub-contractors would prefer to regard the information in the latter form, as they work from a batch of fittings to the positions in which they are fitted, whereas the architect is given a location and has to specify the appropriate item. Reordered schedules can also be useful in the later stages of design when checking, costing and rationalising.

Various forms of output may be possible, but one useful one can be a standard letter facility. If names and addresses are stored on the database, some systems can take the basic text of a letter, insert information from the database into it and print it as often as necessary. So if all clients are to be circulated with a letter, it can be set up and associated with the client database. The program will take each record, extract the client name and address and the name of the contact person and print

Figure 5.5 A KWIC index

the letter with the unique information. It is usually also possible to print the addresses onto adhesive labels at the same time.

Another class of operations that is part of a database management system is the enquiry facilities. Manual schedules have always suffered from the problem that interrogation can be difficult. Queries that match the order of the schedules are not difficult to answer. Given an equipment schedule, for instance, a query such as: 'Is there a telephone in room 1234?' can be easily answered, but a question such as: 'Which rooms have telephones?' will take a long time to answer because there is no index to this information. Not only will a lot of time be spent on such queries, but they are particularly subject to error as it is easy for the eye to miss information, especially when faced with a great deal of data.

The computer can scan a database very quickly and with perfect accuracy, even when given very complicated questions. There are many occasions in the lifetime of a design when enquiries must be made. At the early stages consistency over the whole project can be achieved by cross-checking rooms. Later, odd checks might be made for various reasons. The location of certain items or certain facilities can be checked. In the final stages, site supervision is made easier by being able to answer general queries.

Enquiries can be of almost unlimited complexity, provided that the information is present. So it might be possible when checking fire precautions to extract from a door schedule all doors below a certain width that serve occupied rooms with only one entrance. The three constraints involved could be given to the database system and a list of the rooms satisfying the criteria would be produced. This shortened list could then be considered by the designer to see if the precautions are adequate.

It is of course necessary that the information to answer the query is present. For example, if the architect wanted to check on a window schedule all windows that faced south, this might not be possible because the orientation was not recorded at the time the database was constructed. One of the most difficult problems when setting up a database is to decide what fields should be present. Too many will involve a lot of data entry for little return, while too few will limit enquiries.

In real-life situations the initial recording of the data represents only a small amount of the effort expended. Most of the time is taken up in changing the information as the job progresses and the design is refined. It is this editing of the information that forms the fourth class of operation provided on a database management system.

All systems allow individual records to be referenced and the data in the fields changed as required, but making such one-off changes will probably be rather slower than the equivalent manual process. The computer offers advantages provided there is some sort of system to the alterations because the machine can use its basic ability to carry out the same sequence of operations again and again. The process is conceptually very similar to the enquiry system in that records are extracted that satisfy given criteria, but a further step is then taken in which fields in the records extracted are set to given values. For example, if in a room finishes database the floor finish in all corridors is to be changed to a particular finish, say vinyl tiles, this will be done as follows. The system will check the room name fields in each record for the word 'corridor', then when it finds a match it will set the value in the floor finish field to the text 'vinyl tiles', overwriting the previous value.

Similarly, if a door or window type is to be changed throughout the building, extra filing cabinets put in offices, or any other consistent change made, it can be done much more easily with the help of the computer. As with enquiries, the criteria for extraction can be as complicated as required. Many of the changes to a design will have some consistency, so a lot of time can be saved over making changes by manual means. It can be especially useful near the end of a design when rationalising or changing the standard of equipment finishes throughout the building to save money.

An example might be if the architect were to say: 'We must cut the cost of these finishes. Let's change the oil paint to emulsion paint on all the walls. That goes for the polyurethane paint too, so long as it's on a plastered wall, emulsion looks terrible on concrete. Oh, but we'll leave out the offices; we don't want people to have to stare at a dull finish all day.' This could be translated formally into a database search made up of three parts. The first part would check if the wall-finish field contained the value 'oil paint'. The second part would check the wall-finish value was 'polyurethane paint' at the same time as the wall material field contained 'plaster'. The third part checks for the room-name field containing a value other than 'office'. A combination would then be made of the parts for each record, and the record would be selected for alteration if either one or the other of the first parts was true and the third part were true.

Database systems vary as to the complexity of operations that can be carried out, and in fact some systems would not be able to handle a process as involved as the one described, whereas others could go further and check, for instance, if the word 'office' appeared in any part of the room-name field.

A problem that is encountered when using computer database systems is that absolute accuracy and consistency is essential when entering textual data. When the computer does a search it looks for a particular sequence of characters and the match must be exact. For example, if it was to be asked to look for the word 'corridor' it will not accept 'Corridor': the changing of the first character to upper case prevents the text equating. Any other trivial discrepancy that would be automatically ignored when working manually will cause serious problems on a computer database system. Minor spelling mistakes, reordering words as in 'acoustic tiles' and 'tiles, acoustic' or even putting two spaces between a pair of words rather than one will prevent the correct working of searches.

One way to ensure consistency is to regularly print out individual fields sorted into alphabetical order. Many mistakes will then become obvious, but checking such lists is a tedious business. Another popular approach is to use a short code to represent a longer description. In certain cases, in fact, coding is virtually essential: in an equipment schedule, for instance, a data item such as: 'sink, stainless steel 1,800 mm(W) 600 mm(D), left-hand drainer' is almost impossible to type more than once consistently and would be better replaced by the manufacturer's model number.

Codes are faster and more precise than long descriptions. They can also be made self-checking. Commercially, this is done by adding a 'check digit' to the end of the code. The check digit, which may also be a letter, is derived by a suitable mathematical process from the other characters in the code and if the process is properly chosen will detect over 90% of typing errors. Check digits may be produced and verified by the database system or it may be necessary to use a separate program.

The disadvantage of coding is of course that it is not immediately intelligible. There must therefore be a list of equivalents which can substituted for the code for checking when printing a full schedule. With a relational database this catalogue would be a separate file containing records consisting of each code and its equivalent and substitution is straightforward.

A problem with a catalogue longer than about a hundred entries is that considerable time can be spent in searching through it to find the item required and its codes. The entries may be organised so that the most significant word is first, so a lockable steel cupboard is written as 'cupboard, steel, lockable', and sorted in alphabetical order. This method breaks down when the description contains more than one word that could be regarded as significant. For instance, it is not obvious whether an acoustic hood for a public telephone should be looked up under 'acoustic', 'hood' or 'telephone'.

A popular way of laying out a catalogue that deals automatically with the above problem is the use of a 'Key Word In Context' or KWIC index. In this type of presentation each description is expressed in a natural manner, but is repeated and shifted on the line so that each word in turn is centred, omitting conjunctions and articles. The central words are sorted into order so that it is then easy to scan down the list and pick out an item on the basis of any of its key words. *Figure 5.5* shows a short example. Programs are commonly available to generate KWIC indexes.

Security and privacy

One of the most important problems with databases is the security of the data, that is, its vulnerability to being corrupted or destroyed[56]. Manual methods are fairly secure, the greatest danger being that someone will discard a drawing or vital piece of paper by mistake, but the damage of data in a computer storage system is a regular occurrence and must be guarded against. Damage can occur in several ways: by simple error, as when a file is erased or overwritten by mistake; by program fault, where the user's program or the computer's operating system overwrites a file; by a mechanical fault, where the storage system fails to record data or even physically damages the storage medium or by deterioration of the data, as when the signal on a magnetic tape fades with time.

Once data has been lost, even if a paper record of it can be found, the effort and time involved in re-entering it into the computer will very great. It is therefore essential to make provision for loss. This is most effectively done by taking an extra copy of the database file. Depending on the amount of updating that takes place, this may be done every day, or perhaps once or twice a week. Making a copy takes only a few minutes of computer time, so with little difficulty data can be safeguarded. The discs or tape containing the copies should be stored in a fireproof safe, or if this is not available, in a location well away from the computer itself so that an outbreak of fire will be unlikely to damage both. The possibility of malicious damage must also be considered and again requires secure storage for the copies.

An associated problem that has received a lot of publicity is the privacy of computer data[57]. As a computer database can be copied so easily, it is feared that unauthorised persons could readily obtain information they should not have and as the database can be searched for specific information quickly and cheaply, violation of privacy could take place on a large scale. This has not previously been a problem simply because copying the contents of a conventional filing cabinet would take a long time and be very obvious and also because most of the information contained is not cross-indexed and so is not readily accessible.

Privacy is usually guarded on a database system by means of passwords. It is normally necessary to enter a password to access a computer at all, the operating system having a list of authorised alternatives. Most database systems, although by no means all, also offer password protection before a file can be accessed. A few systems go further and can have several levels of password protection, thus allowing certain users access to only some of the fields or allowing them to read data but not to alter it.

While privacy is a serious problem, it is not in general relevant to an architect's databases because the information collected is not particularly sensitive. It would be unusual, for example, for an architect to care who reads the door schedules; the problem is much more likely to be to find someone prepared to check them. However, occasionally information will be sensitive. If the computer is being used for accounting, especially payroll applications, some protection is necessary. Also if some building is being done for a defence establishment, for example, the names of the rooms could give away the function and status of the building and this might be undesirable. In these circumstances it will be necessary to buy a database system that has password protection, or to lock away the computer when it is not in use.

Reference databases

A reference database is a database that contains a complete set of information on a given subject which the user can search for the data required in the same way as he might look through reference books, catalogues and so on. The advantage over the manual alternatives is principally the greater speed of access and more flexible ways of searching for information.

The idea of setting up and making generally available reference databases has always been a popular one and many projects have been launched since the mid-1960s. Most of these projects failed after a comparatively short period and the reason appears to be that the setting up of a completely comprehensive database on almost any any subject is a long and expensive business and the effort of keeping the information up to date is also an expensive and continuing task. It must also be said that printed books have advantages over computers, such as accessibility, familiarity, and most importantly their ability to include high-quality diagrams and photographs, and these factors hinder the acceptance of computer alternatives.

Many examples might be given of projects which were closed down or curtailed after a comparatively short period, but some of the most ambitious of these were set up in Britain.

In 1968 the RIBA began a project to produce a library of standard clauses for use on job specification. Development was formally halted in 1976 after over £180,000 had been spent[58]. However, the project was advanced enough to continue under its own impetus and is now in established use.

In 1979, the National Building Agency launched an information system for the construction industry under the name of Contel[59]. The information offered included technical references, cost information on building materials, guides to design criteria and news items. However, the service was not immediately popular, and a cutback on funding two years later caused its demise.

Other examples might be given, but it should be clear that if publicly-available databases of this type are not viable, then private databases operated within a single firm are unlikely to be.

Reference databases are still available but tend to be less ambitious in the scope of what they offer. An internationally popular system is offered by Pergamon Infoline, a subsidiary company of the Pergamon Press[60]. This firm supports a range of databases which contain references to current technical literature. The references are taken from journals, books, conference papers, government publications and so on. Databases are available for many disciplines, but the ones of primary interest to architects are PICA, which holds references to papers in construction and architecture, and IBSEDEX, which covers mechanical and electrical services in buildings. The topics covered in PICA encompass energy conservation, health and safety in building, building failures, interior design and many other subjects. The references extend back to 1974 and at the time of writing the database contains 45,000 entries. IBSEDEX is a more specific database and covers the whole area of mechanical and electrical services including heating and cooling, air-conditioning, lighting, noise, corrosion and fire protection. Its references extend back to 1979 and are made up of 15,000 records.

Databases concerned with building performance are run by the Architectural and Engineering Performance Information Center (AEPIC) at the University of Maryland, USA, and by the Construction Performance Centre in Britain. These organisations collaborate in the exchange of information[61].

There are many other databases in existence, most of which were created originally by large organisations for their own use and are now also available to the public[62].

Most services are accessed by means of a terminal or microcomputer over the telephone lines. A popular alternative with the smaller-scale services is to distribute regularly magnetic discs or tapes containing the latest version of the database.

Another way of transferring information is provided by viewdata methods. These also send information over telephone lines but display it on a domestic television that has been fitted with decoding circuitry, which is available quite cheaply. Viewdata is easier to use as it is menu-driven with the user being given a fixed list of choices at each point, and also has the capability of displaying rather crude graphics. It is however slower and less flexible in use.

The electronic publishing firm of Context Online Ltd. is currently providing a wide range of information to architects through a private viewdata system. This service is known as CONTEXT. It has many aspects, but among the most useful are a guide to over 8,000 suppliers, pricing information on over 20,000 products, references to technical literature, building regulations and construction industry standards and a group of on-line computer programs. This last feature is especially interesting: it means that an architect can run programs through the viewdata service without actually owning a computer. *Figure 5.6* is a page from the index of suppliers showing a typical product description.

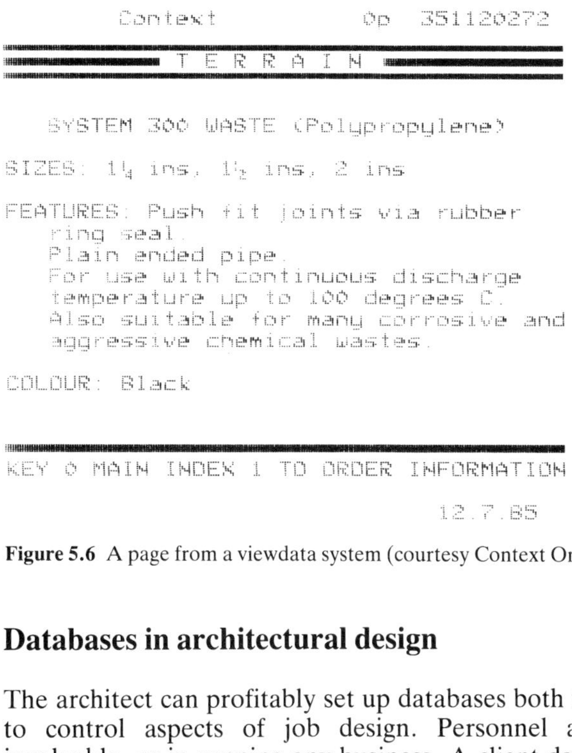

```
        Context              Op   351120272


                    T E R R A I N


    SYSTEM 300 WASTE (Polypropylene)

SIZES: 1¼ ins, 1½ ins, 2 ins

FEATURES: Push fit joints via rubber
          ring seal.
          Plain ended pipe.
          For use with continuous discharge
          temperature up to 100 degrees C.
          Also suitable for many corrosive and
          aggressive chemical wastes.

COLOUR: Black



KEY 0 MAIN INDEX 1 TO ORDER INFORMATION

                          12.7.85
```

Figure 5.6 A page from a viewdata system (courtesy Context Online Ltd)

Databases in architectural design

The architect can profitably set up databases both for administrative purposes and to control aspects of job design. Personnel and accounting databases are invaluable, as in running any business. A client database can be used to aid client relations and also to analyse the performance and main activity areas of the practice over a preceding period.

On a specific job, there is a number of areas in which written schedules are necessary or helpful in coordination. Door, window, interior finishes and fittings schedules are most commonly prepared, but others may be useful depending on the job in hand[63]. The most obvious approach might seem to be to hold all the information on each job in a single database of the hierarchical or relational type. In practice, however, it is usually better to split the job into separate databases for each subject. This is because a single large database will often exceed the disc capacity of a microcomputer and will in any case be slow and cumbersome to use.

It is also true that the design of a building is an evolving process and the focus of attention changes from one subject to another with time. At the start of the job, very basic information is being built up on which rooms will be present and what their physical requirements will be. At a later stage, this sort of information is frozen and incorporated into the sketch design, but detailed information on doors and windows is being built up. Later still, the designer will decide on the room finishes and check manufacturers' catalogues for the exact fittings to be installed.

One of the first databases that can be created during the design of the building is that for room-briefing information. This database will hold the requirements from the client's brief and will be extended as the architect and the consultants generate basic information on every room or activity area in the building. Initially, this information might consist of only the most basic facts such as the use, floor area and occupancy of the rooms, but later it can be refined to include environmental data, the services supplied, the finishes and so on. This data can be used at the feasibility and outline proposal stages of the design to agree standards with the client and specialist consultants and later as a check list when designing the rooms in detail.

A convenient way to collect the data is for the architect to produce a set of skeleton data sheets, each containing the basic information on one room, which is circulated to all the members of the design team. In this way, most of the information that the designer needs can be gathered quickly and with consistency. A typical sheet with handwritten additions is shown in *Figure 5.7*.

```
LOAMSHIRE GENERAL HOSPITAL - ROOM BRIEFING INFORMATION
```

```
01.ROOM NAME                        ANAESTHETIC ROOM
02.ROOM NUMBER                      5013
03.FUNCTION                         - Induction of anaesthesia and hypothermia.
04.SIZE (SQ.M.)                     - 38 ½
05.OCCUPANTS                        - 4.
06.PERIODS OF USE                   - 24 hrs
07.FUNCTIONAL RELATIONSHIPS         - Adjacent operating theatres.
08.SPECIAL CHARACTERISTICS          -
09.TEMPERATURE                      - 24 - 27°C  50% RH (adjustable)
10.VENTILATION                      - Supply & exhaust with high efficiency filters, 100% FA +ve pressure, NC 30 (max)
11.ILLUMINATION                     - 100 lux (general) + special lighting
12.LIGHTING                         - 2 wall mounted anaesthetic lamps
13.DOORS                            - To admit bed. Single swing door with 90° check. Vision panel.
14.WINDOWS                          - Automatic floor.  A very high standard of cleanliness is required
15.SPECIAL FINISHES                 - S/S sink.
16.SANITARY FITTINGS                - H + C
17.WATER SERVICES                   -
18.SPECIAL WASTES                   - oxygen, Nitrous oxide. Suction
19.PIPED SERVICES                   - For anaesthetic lamps + clock. Twin sockets
20.ELECTRICAL SUPPLIES              - -
21.SPECIAL ELECTRICS                - -
22.TELEPHONES                       - -
23.CALL SYSTEMS                     - -
24.CLOCKS                           - Sync with sweep seconds hand
25.FITTINGS                         - -
26.EQUIPMENT                        - -
27.MISCELLANEOUS                    - -
```

Figure 5.7 A room-briefing schedule with handwritten amendments

These sheets form a valuable design in themselves, but entering the information into a computer database gives the advantages of better communication, easier amendments and interrogation facilities. A typical sheet from a computer-held room-briefing database at an advanced stage in the job is shown in *Figure 5.8*. The system used gives the ability to have multiple entries in certain fields.

The information from individual fields can be printed out and used for rough pricing and for consistency checks. The designer will then use this refined data in creating the outline design. The act of drawing out the design will inevitably show up practical problems and lead to changes being made; it is most convenient if these changes are recorded on the database.

At a later stage, the door and window schedules will need to be prepared, and here the room-briefing database can be used as a basis. Most database systems will

```
LOAMSHIRE GENERAL HOSPITAL - ROOM BRIEFING INFORMATION


01.ROOM NAME                    ANAESTHETIC ROOM
02.ROOM NUMBER                  5013
03.FUNCTION                     INDUCTION OF ANAESTHESIA AND HYPOTHERMIA
04.SIZE (SQ.M.)                 38.5
05.OCCUPANTS                    4
06.PERIODS OF USE               24 HOURS
07.FUNCTIONAL RELATIONSHIPS     ADJACENT OPERATING THEATRES
08.SPECIAL CHARACTERISTICS      -
09.TEMPERATURE                  24-27 DEG.C.
                                50% RH (ADJUSTABLE)
10.VENTILATION                  SUPPLY/EXHAUST HIGH EFF.FILTERS,100% FA POS.PRESSURE,NC 30(MAX)
11.ILLUMINATION                 100 LUX (GENERAL)
                                SPECIAL LIGHTING
12.LIGHTING                     WALL MOUNTED ANAESTHETIC LAMP,2 NO.
13.DOORS                        SINGLE SWING,VISION PANELS,90 DEG. CHECK,TO TAKE BED
14.WINDOWS                      -
15.SPECIAL FINISHES             ANTI-STATIC FLOOR
                                A VERY HIGH STANDARD OF CLINICAL CLEANLINESS
16.SANITARY FITTINGS            STAINLESS STEEL SINK
17.WATER SERVICES               HOT WATER
                                COLD WATER
18.SPECIAL WASTES               -
19.PIPED SERVICES               OXYGEN
                                NITROUS OXIDE
                                SUCTION
20.ELECTRICAL SUPPLIES          TWIN SOCKETS
                                TO ANAESTHETIC LAMPS
                                TO CLOCK
21.SPECIAL ELECTRICS            -
22.TELEPHONES                   -
23.CALL SYSTEMS                 -
24.CLOCKS                       SYNCHRONOUS,WITH SWEEP SECONDS HAND
25.FITTINGS                     -
26.EQUIPMENT                    -
27.MISCELLANEOUS                -
```

Figure 5.8 A completed room-briefing schedule

allow a single field to be output and reformatted into a separate database file. If this is done with the fields for doors and windows in the room-briefing database, an accurate skeleton database can be created in a few minutes. It is then necessary to go through, say, the door schedule door by door and insert more precise data, but because information can be repeated where two doors are the same, this need not take long in a rationalised design.

At different times, this process can be repeated for other subjects such as furnishings, fittings, sanitary-ware, room finishes and so on. The structure of the new databases will vary according to the information held, and it is a good idea to keep the structure as simple as possible. Door, window and finishes databases can normally be held in a simple tabular form, with single entries under each heading. The page from a typical interior finishes database in *Figure 5.9* illustrates this. Contract fittings or equipment databases will have a more complex structure with multiple entries in most rooms.

Because databases can be set up relatively easily, it is often useful to create them and produce formal printed schedules where it would be more usual with manual methods to rely on the drawings. When ordering, pricing or supervising on site it is faster and less error-prone to use a printed list rather than a drawing.

The SUPERFILE database management system

SUPERFILE is a relational database management system written by Southdata Ltd. It is implemented on the IBM PC and compatible machines and also on the Apricot and Sirius microcomputers. Currently, it is used at over 2,000 sites.

An important characteristic of SUPERFILE is that it is very flexible. Most database systems impose rigid constraints on the database structure and limitations

LOAMSHIRE GENERAL HOSPITAL   ROOM FINISHES   DATE: 30/04/86

ROOM	ROOM NAME	FLOOR	SKIRTING	WALL	WALL FINISH	CEILING	CEILING FINISH
2201	STAIR (MAIN)	PVC WELDED SHEET	HARDWOOD	PLASTER	EMULSION PAINT	PLASTER	EMULSION PAINT
2202	ENTRANCE HALL	CARPET	-	FACING BRICK	SELF FINISH	TIMBER SLATTED	SEALED
2203	DUCT	CONCRETE		FACING BRICK	SELF FINISH	CONCRETE	SELF FINISH
2204	RECEPTION	CARPET	PVC 100MM PLAIN	PLASTER	EMULSION PAINT	PERFORATED METAL	SELF FINISH
2205	RECEPTION COUNTER	CARPET	PVC 100MM PLAIN	PLASTER	EMULSION PAINT	PERFORATED METAL	SELF FINISH
2206	OFFICE - RECORDS	CARPET	PVC 100MM PLAIN	PLASTER	EMULSION PAINT	PLASTER	EMULSION PAINT
2207	OFFICE - NURSING STAFF	CARPET	PVC 100MM PLAIN	PLASTER	EMULSION PAINT	PLASTER	EMULSION PAINT
2208	WAITING AREA	CARPET	PVC 100MM PLAIN	PLASTER	EMULSION PAINT	PERFORATED METAL	SELF FINISH
2209	TREATMENT ROOM	PVC WELDED SHEET	PVC 100MM COVED	PLASTER	OIL PAINT	PLASTER	OIL PAINT
2210	DISPOSAL ROOM	PVC WELDED SHEET	PVC 100MM COVED	PLASTER	OIL PAINT	PLASTER	OIL PAINT
2211	CUPBOARD - ELECTRICAL	CEMENT	-	FACING BRICK	EMULSION PAINT	CONCRETE	EMULSION PAINT
2212	DUCT	CONCRETE	-	FACING BRICK	SELF FINISH	CHEQUER PLATE	MATT PAINT
2213	TROLLEY PARKING	VINYL TILES	PVC 100MM COVED	PLASTER/TILES	OIL PAINT	ACOUSTIC PLASTER	EMULSION PAINT
2214	PANTRY	VINYL TILES	PVC 100MM COVED	PLASTER/TILES	OIL PAINT	PLASTER	OIL PAINT
2215	WASH AREA	VINYL TILES	PVC 100MM COVED	PLASTER/TILES	OIL PAINT	ACOUSTIC PLASTER	EMULSION PAINT
2216	TELEPHONE	CARPET	PVC 100MM PLAIN	PLASTER	EMULSION PAINT	PERFORATED METAL	SELF FINISH
2217	SERVERY	VINYL TILES	PVC 100MM COVED	PLASTER/TILES	OIL PAINT	ACOUSTIC PLASTER	EMULSION PAINT
2218	DINING ROOM	PVC WELDED SHEET	PVC 100MM COVED	PLASTER	EMULSION PAINT	PERFORATED METAL	SELF FINISH
2219	CLOAKROOM - PATIENTS	PVC WELDED SHEET	PVC 100MM COVED	PLASTER	EMULSION PAINT	PERFORATED METAL	SELF FINISH
2220	CLEANERS ROOM	PVC WELDED SHEET	PVC 100MM COVED	PLASTER	OIL PAINT	PLASTER	OIL PAINT
2221	CORRIDOR	CARPET	PVC 100MM PLAIN	PLASTER	EMULSION PAINT	PERFORATED METAL	SELF FINISH
2222	CUPBOARD - ELECTRICAL	CEMENT	-	FACING BRICK	EMULSION PAINT	CONCRETE	EMULSION PAINT
2223	LOBBY	PVC WELDED SHEET	PVC 100MM COVED	PLASTER	EMULSION PAINT	PERFORATED METAL	SELF FINISH
2224	STAIR (SECONDARY)	PVC WELDED SHEET	HARDWOOD	PLASTER	EMULSION PAINT	PLASTER	EMULSION PAINT
2225	STORE - OCC.THERAPY	PVC WELDED SHEET	PVC 100MM COVED	PLASTER	EMULSION PAINT	PLASTER	EMULSION PAINT
2226	STORE - OCC.THERAPY	PVC WELDED SHEET	PVC 100MM COVED	PLASTER	EMULSION PAINT	PLASTER	EMULSION PAINT
2227	HOSE REEL RECESS	ETERNIT	-	PLASTER	EMULSION PAINT	PLASTER	EMULSION PAINT
2228	THERAPY AREA - GROUP	PVC WELDED SHEET	PVC 100MM COVED	PLASTER	EMULSION PAINT	PERFORATED METAL	SELF FINISH
2229	OFFICE - THERAPIST	CARPET	PVC 100MM PLAIN	PLASTER	EMULSION PAINT	PLASTER	EMULSION PAINT
2230	OFFICE - GENERAL	CARPET	PVC 100MM PLAIN	PLASTER	EMULSION PAINT	PLASTER	EMULSION PAINT
2231	THERAPY AREA - CRAFTS	PVC WELDED SHEET	PVC 100MM COVED	PLASTER	EMULSION PAINT	PERFORATED METAL	SELF FINISH
2232	STAIR (ESCAPE)	PAINTED METAL	-	-	-	-	-
2233	THERAPY ROOM (CLERICAL)	PVC WELDED SHEET	PVC 100MM COVED	PLASTER	EMULSION PAINT	PERFORATED METAL	SELF FINISH
2234	LOBBY	WOODBLOCK	PVC 100MM COVED	PLASTER	EMULSION PAINT	PERFORATED METAL	SELF FINISH
2235	THERAPY ROOM (WORKSHOP)	WOODBLOCK	HARDWOOD	FACING BRICK	EMULSION PAINT	PERFORATED METAL	SELF FINISH
2236	OFFICE - TECHNICIAN	WOODBLOCK	HARDWOOD	PLASTER	EMULSION PAINT	PERFORATED METAL	SELF FINISH
2237	STORE	WOODBLOCK	HARDWOOD	PLASTER	EMULSION PAINT	PLASTER	EMULSION PAINT
2238	STORE - TIMBER	GRANTOP	-	FACING BRICK	EMULSION PAINT	CONCRETE	EMULSION PAINT
2239	THERAPY - REHAB.KITCHEN	VINYL TILES	PVC 100MM COVED	PLASTER/TILES	OIL PAINT	PLASTER	OIL PAINT
2240	LAVATORY - FEM.PATIENTS	VINYL TILES	PVC 100MM COVED	PLASTER	OIL PAINT	PLASTER	OIL PAINT

Figure 5.9 An output from a tabular database

on the form that the data may take. SUPERFILE has far fewer constraints. This can be especially useful in architectural applications where information is often vague or lacking at the beginning of a design and is defined later.

The system can allow several users to access the same database simultaneously. This feature can be useful when working on larger projects as designers at a number of terminals can look at the same information. In order to prevent contradictory situations arising, SUPERFILE imposes limitations on access in certain cases.

The system consists of a suite of interlocking programs. The core program is also called 'Superfile' and handles the basic storage and retrieval of information. When necessary, this is invoked by the program 'Supertabs', which deals with the printing of reports and by the program 'Superforms', which sets up screen displays for handling data by the form-fill technique. There is also a number of small utility programs to carry out various functions. A SUPERFILE database is held in a single disc file. There are no limits on size; the database may grow to the maximum available disc capacity. There is little formal definition of the structure of the database. The user can initially define the names of some fields (or 'tags' as they are called in SUPERFILE terminology) and input data under those headings. However, if at some future time it were necessary to add a field, this could be done with a single command and it is then possible to add information to any record under that heading. Most database systems require the file to be completely rebuilt before a field can be added.

Flexibility is also available in the way in which the information can be supplied. Any field can hold any data: it is not necessary, for instance, to specify that only numbers be acceptable under a given heading. This is a convenient facility where information is lacking. For example, if recording the cost of a piece of equipment, it is possible to enter the value 'to be ascertained' and later replace it with a firm price. Also, no restrictions are placed on the length of the data value supplied. A text string can be as long as necessary and a number can have as many significant figures as necessary. Space will be allocated in the database as required.

For efficiency reasons, most database systems allocate a fixed amount of space for each record. When creating such a database the number of fields and the maximum size of the fields must be defined, and when a record is created this amount of space is reserved in the disc file and all subsequent operations on the record will reference that space. This is a very fast method as the precise position of every field within every record can be found directly from the record number. SUPERFILE stores data in a far more flexible manner, but the price it pays for having this flexibility while retaining speed is that cross-reference indexes are set up on every field and in the worst case these can occupy as much space on disc as the data itself.

More than one data value can be placed in a single field of a single record. This should not be regarded as any sort of hierarchical system as it is not possible to perform processing on multiple values; the facility is intended to be used when a single data item has several parts. For example, a field for remarks might contain a single sentence made up of several lines.

SUPERFILE can be used as a simple tabular database for straightforward applications. For more complex forms of data organisation it will typically be used in a relational way. This implies that effectively there is more than one type of record present with common fields to provide a cross-reference. As an example, we could consider a room equipment database. This could be set up by declaring the fields ROOMNAME, ROOMNUMBER, ITEM and QUANTITY. For each room, the user enters a record containing entries under ROOMNAME and ROOMNUMBER, then for each different item of equipment enters a record containing entries under ROOMNUMBER, ITEM and QUANTITY. The room number entry will link these two types of record together.

There is no practical limit on the number of fields that may be present and no penalty in leaving fields blank in some records, so it is possible to set up very complicated forms of data organisation. If in the equipment database, for example, a field was added named SUPPLIER it would be possible to enter records containing entries under ITEM and SUPPLIER. There would then be a link between each room, the equipment within it and the firms that supply each piece of equipment.

SUPERFILE provides a number of sophisticated ways of searching for data. It is of course possible to check for simple equality. The command NAME = SMITH might be used on a personnel file. Comparisons on a value being greater than or less than a test value can be made using the appropriate mathematical symbols. Thus: SALARY > 10000 or: SERVICE < 10 would also be typical searches performed on a personnel database. Unusually, it is possible to perform searches with incomplete information. The most striking method of doing this is to use the '@' sign (chosen because of its resemblance to a human ear) to search for text that sounds like a given word. So for example a search of the form: NAME @ AUSTEN will find not only AUSTEN, but also AUSTIN, ASHTON and ASHTEN.

The character '?' can be used to indicate that any character can be substituted when making the comparison. In this way NAME = S???H will find say, SINGH, SMITH and SOUTH. The character '★' indicates that any number of any characters can be substituted in the comparison. So NAME = ★SON will find, say, JOHNSON, ROBINSON and SMITHSON. Various other text-matching facilities exist, but although ingenious they would probably be of only occasional use to architects.

The system aims to be easy to use and to prompt the user as far as possible. There are two main ways in which it does this. First by offering menus of fixed choices when decisions have to be made and second by collecting data by displaying a 'form' on screen which the user fills in.

The form-fill technique is implemented by the program displaying a number of headings under which information must be given. The headings may be arranged in any pattern over the screen to give the greatest clarity. A space is provided alongside each heading into which data must be typed. The user is not able to type outside the spaces. With most methods of data input, common mistakes are to omit items of data or to enter items in the wrong order. The form-fill technique makes such errors less likely. *Figure 5.10* shows a typical screen form.

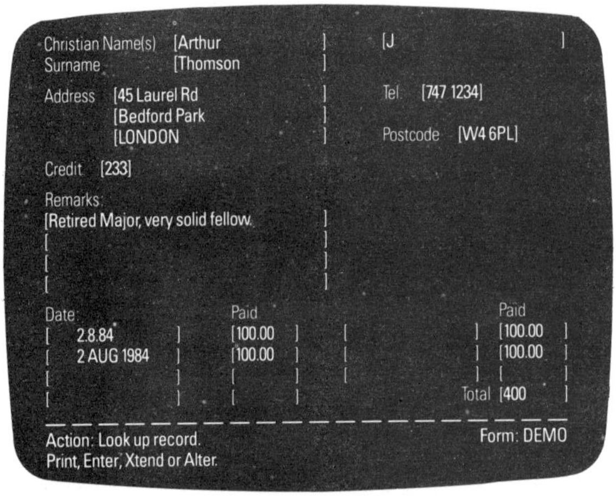

Figure 5.10 A SUPERFILE screen form (courtesy Southdata Ltd)

Designing a form is done in a natural manner and involves two stages; the first stage to define the layout and the second to define the properties of the form. In the first stage the user starts with a blank screen and types in relevant prompts, remarks, titles and so on at any location. At this stage also, the user specifies the spaces on screen that will be filled with data from a record being considered. These are indicated by simply typing square brackets at either end of the space.

The second stage of form definition is concerned with specifying which fields from the records will be considered and which properties will be associated with them. Each space delimited by square brackets is highlighted in turn and the user must specify the name of the field whose data is to occupy the space.

The user can go on if necessary to select from various options that work on the data. The options can be broadly divided into three types: those for searching, those for checking data input and those for processing data.

The searching options are used when a form will be used to look for records having specific data. They have the same layout as has already been described and use mathematical symbols to perform comparisons. If several conditions are specified on a form, all of them must be met for a record to be selected. It is not possible to select a record on the basis that one or the other condition is fulfilled.

The data checking options ensure that on data input the form will check information typed into the bracketed spaces. It is possible to specify that data may not be entered at all in some fields of a particular form, thus protecting vital information from accident overwriting. It is possible to check that numbers contain a maximum number of digits and decimal places or that numbers fall inside a specified range. If dates are to be entered within a field, their validity, in several different common forms, can be checked. A less common, but very useful, checking option is against a list of possible entries. Thus if recording the sex of an employee only 'M' or 'F' could be made acceptable. An option exists to check that information placed in a field is unique. So a field containing the room number could be made unique and will prevent the same number being used for two different rooms. Various other checks are possible and can greatly reduce the error rate in data input.

The data processing options enable fields to be filled automatically with the results of a calculation or other process. The simplest, but one of the most useful, options is the 'stayput' facility, which specifies that a field is to be filled with a standard value. When the form is used for data input, that value can be overtyped or cancelled as required. When data input is repetitive, this option can save a lot of typing.

More complicated processing is made possible by the 'calculate' option. When applied to a field, data can be taken from other fields on the form, mathematical operations performed and the result stored. Thus if a job accounting database contained fields for total fees received, various costs incurred, and profit, the profit field could be filled automatically. If the user then required to check the jobs making a loss, the database could be searched with a form that selected negative values in the profit field. This is a way of circumventing the restriction that it is not possible to search by comparing values in one field with values in another.

When a form has been created, it can be stored and subsequently invoked for use. There are four main operations that can be performed on records through a form; these are searching, adding, deleting and altering. Obviously, a record must be located before it can be deleted or altered, so these two operations also involve searching. Operations are not performed blindly as a completely automated process, but rather each record must be checked individually by the user. This is rather slower than the automated method, but far safer as mistakes cannot corrupt or destroy the database.

To perform a search, the user must give the name of a form, or may set one up on the spot for a one-off enquiry. Each record meeting the conditions is then displayed in turn on the screen in the format dictated by the form. The user can print that record for reference if required and go on to the next record by pressing carriage return.

The deletion command searches in the same manner, but when each record is displayed the user is asked to confirm that it is to be removed. The answer 'Y' or 'N' will have the appropriate effect and the search will continue.

The alteration command again searches in the same manner. When a record is found the user is asked to confirm that it is to be altered. If the answer is affirmative, each bracketed space is taken in turn and new information can be overtyped as required. After the last field, the changed record can be stored and printed if required, or further changes may be made if a mistake was made in typing. On storing, any calculation fields will be activated and the results will overwrite the previous contents of those fields.

Adding records requires the user to skip through each bracketed space, typing in information, except when the 'stayput' or calculation options are in force, in which case the data will be generated automatically. The data typed will be verified automatically when checking options have been set. After data for the last field has been entered, the record can be stored and printed if required or changes made to the information supplied. The program is then ready to add another record through the same form.

Printing reports is done through the separate 'Supertabs' program. The program is divided into two parts. One part is concerned with creating a kind of form called a Report Definition which will dictate how the report is to be laid out and the other part applies the definition to the database. The definition can be associated with a form generated by Superforms which will extract only certain records from the database. If the report is to be sorted on some field, as would normally be required, the entire database must be sorted by a separate program before entering Supertabs.

The creation of a Report Definition is broadly similar to the creation of a form. Titles and comments can be inserted as required. The user must define 'Report Lines' which control the way in which each record is printed, and also lines to control the printing of totals and sub-totals where these will be required. Report Lines can again include notes and comments and special symbols are used in the same way as square brackets in Superforms to define spaces where field data will be printed. As in Superforms, it is possible to associate options with these spaces. The principal options are concerned with performing automatic calculations and totalling.

It can be seen that the SUPERFILE system is extremely flexible in the way in which it can store data, but has some limitations in the facilities it provides for manipulating and printing this data. This arises from the stated philosophy of its originators which is that it is pointless to learn a very complex language for use in one particular program, where one of the ordinary high-level languages would do as well. SUPERFILE provides a full linkage to several well-known languages including BASIC, COBOL and PASCAL, and programs may be written in one of these languages to perform tasks of any desired complexity.

Chapter 6

Computer-aided draughting

Principles

Drawings have always been the architect's main means of storing and conveying information, and as the essential information is spatial this must continue to be so. The architect designs by drawing; the engineer uses drawings as the basis for his calculations, the cost engineer or quantity surveyor takes off most of his information from them and the contractor builds from them.

Although computer-aided draughting has been possible for a long time, it is only fairly recently that it has begun to be cost-effective. Graphical applications inherently require a lot of processing power and this has meant that they have been too expensive compared with the benefits that they can provide in architectural practice. Fortunately, the continuous fall in the cost of circuitry has now brought computer-aided draughting well within the bounds of viability[64].

Computer-aided draughting systems can provide equivalents for all the usual manual operations and also have extra capabilities for which there are no manual equivalents. The most important of these is probably their ability to use the same drawing many times in different contexts as parts of a larger drawing. At its simplest level, this eliminates the frequent redrawing of standard building components and fittings. At the the beginning of a job, a range of drawings of standard elements such as columns, doors, fittings and services symbols can be defined or taken from a library of such drawings and copies of those positioned on the main drawings as required. The elements may be as complex as required and may in certain cases consist of hundreds of lines, arcs and alphanumeric characters, but a typical standard element could be a door plan such as the one shown in *Figure 6.1.*

Figure 6.1 A typical graphic element (segment)

The concept of a repeatable drawing element is common to all computer draughting systems. There is no agreed way of referring to such elements and terms such as 'instance', object', 'cell', 'group' and 'component' are in use by different programs; however, the recent international standard for exchanging graphic information, IGES, uses the word 'segment' and this is emerging as the most popular term.

Once defined, a segment can be used in any position on a drawing just by giving its associated code to the computer. The orientation can be changed and the segment can also be mirrored. Thus a left-hand swing door can also be used as a right-hand swing, a sink can have its drainer on the opposite side and so on. *Figure 6.2* shows the door segment used in several different contexts in a portion of a typical plan.

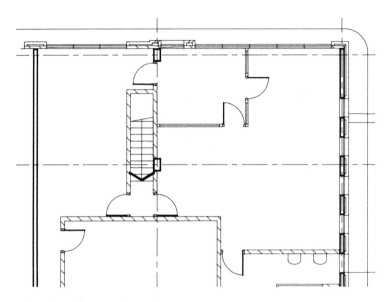

Figure 6.2 The repeated use of a segment

The use of these repeated segments can save a lot of time on a large drawing and can improve the quality of the drawing because the standard can be drawn in detail and very accurately without imposing any time penalty when constructing the main drawing. *Figure 6.3* shows a building elevation in which a segment was created for a single window then duplicated as necessary.

At a higher level, most systems allow copies to be made of groups of previously assembled segments. Thus a standard office layout could be drawn then repeated at a number of locations in a building, or a standard ward plan could be used several times in a hospital project. In an extreme case, if a plan is symmetrical, only half of it need be drawn, then a mirrored copy is made and the two drawings are joined together to form the complete plan. *Figure 6.4* shows an example of the repeated use of a room plan which contains a number of segments including those for the structure, the door, the desk, the telephone etc.

Another very useful capability that cannot adequately be duplicated by manual methods is the overlaying of drawings containing different information to make up various composite drawings. For example, the information to produce a drawing of the wall and structure plan can be stored in the computer separately from the information for furniture, sanitary fittings and service outlets. These files can then

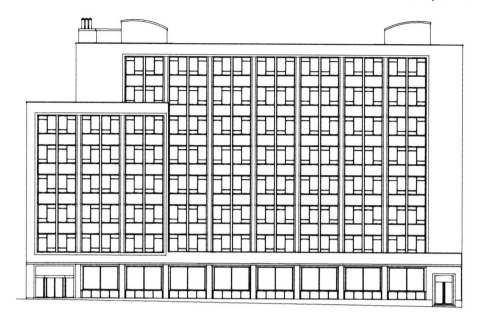

Figure 6.3 The repeated use of a complex segment

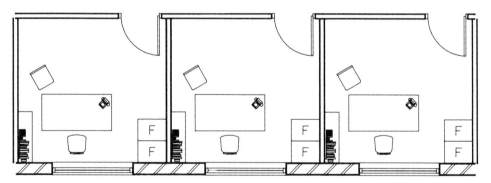

Figure 6.4 The repeated use of a group of segments

be combined in a number of different arrangements to make up drawings for different purposes. The walls and structure by themselves comprise an outline plan to give a clear view of the design for evaluation or for the contractor to build from (*Figure 6.5*). The structure plus the serviced items, such as electrical equipment, sinks and sanitary fittings, gives a drawing for the mechanical and electrical services engineer to work from (*Figure 6.6*). Finally, all the information together in a single drawing gives a floor data plan for a complete view of the design (*Figure 6.7*).

Other overlays might include dimensions, planning grids, ceiling layouts and so on and can be used to produce a wide variety of plans with little effort and can greatly improve communication.

In theory, overlaying is possible with manual methods, but in practice it is little used: multiple layers of tracing paper become opaque, and accurate alignment is

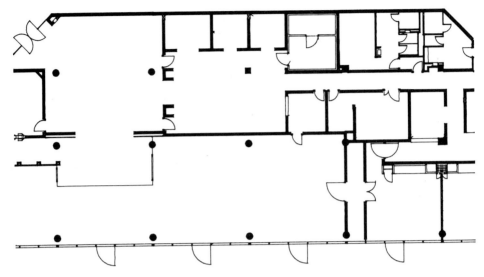

Figure 6.5 A computer-produced structure overlay

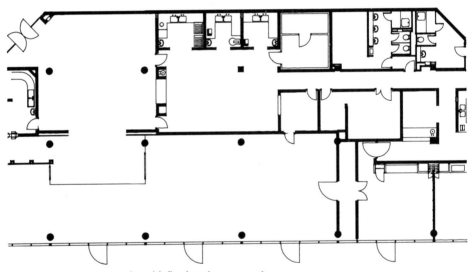

Figure 6.6 A structure overlay with fixed equipment overlay

almost impossible with drawings of any size. The computer, however, can display several drawings on a screen at once, or output several drawings to the plotter on top of one another, with no loss in clarity and with perfect alignment. It is therefore quite practicable to build up drawings as several layers.

Another important advantage of computer drawings is that they are held inside the machine in full-size dimensions and any area of interest can be output on the plotter at any required scale. Thus it is possible to have a plan of an entire floor of a building at convenient scale, 1:100 say, to give an overall impression. It is also possible to plot out large-scale plans, say at 1:20 scale, of single rooms or groups of

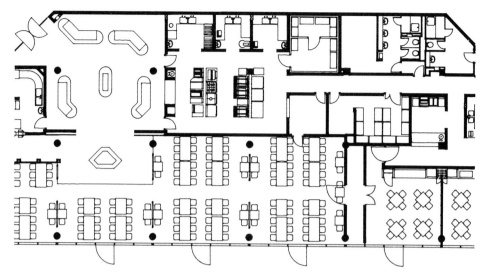

Figure 6.7 A structure overlay with fixed equipment overlay and furniture overlay

rooms to show the detailed fitting-out. Commonly, the extra information necessary on a large scale plan will be held as a separate overlay and plotted out when necessary.

This technique saves a lot of draughting time, because otherwise much of the information would have to be drawn twice at the two different scales. An important additional advantage is that if any changes are made to any aspect of the design, only a single alteration need be made to the drawing that contains that element of information. With manual drawing, if the position of, say, a wall is altered, care must be taken to alter all the other plans that show that wall. This can occupy a lot of time, as typically several sets of plans will have to be kept in step. There is also the possibility of errors creeping in. These problems do not arise with a computer system as, in the example given, only the structural overlay need be altered and the correction will then appear on all subsequent drawings.

Computer-aided draughting systems can also make alterations to drawings faster than is possible with manual methods. Very few drawings are right first time; most undergo change and refinement as the design progresses. Manual erasure of ink lines involves laborious removal with an eraser or scalpel. The result, especially when a fairly large area is erased, is often very messy. If yet another alteration is required in the same place,the result may be a hole in the paper! It is a common occurrence in most offices that a drawing needing extensive modifications is scrapped and redraughted. With a computer-aided system, erasure of single lines is quick and leaves no trace of their former presence. Even quicker erasure is possible if an entire segment, such as a door, is to be removed. A single indication of the segment will result in its elimination from the drawing file and remove all the lines that comprise it. Most systems also provide a facility to erase an entire area of the drawing with a single command if it is being completely replanned.

The increase in speed is yet more dramatic if, instead of being removed completely, a segment is moved to another location. This often happens in replanning, when furniture and items of equipment and perhaps the position of doors and windows may be rearranged in a room. Manually, of course, this requires erasure and complete redrawing of each item of equipment, but with a computer system all that is required is a simple adjustment of the location of the segments and possibly their rotation. On a larger scale, where a late decision might be taken to adjust the relative positioning of two building units on a site, this facility could save days or weeks of drawing.

A common use of the quick editing facility is when two or more floors of a building are similar, but not identical. This is often the case: structural walls, window positions, stairs and lift shafts will be the same, but the internal walls will vary to some extent. In this case, copies can be made of an entire drawing of one floor and specific amendments made to them as necessary for other floors.

At another level, computer editing facilities can be of great help in the use of working details, such as those showing the exact construction of parts of the building. Many architects' offices have attempted to use standard working details for all their projects, but not many have been successful because each job seems to require slightly different dimensions or different materials or extra features which make it impossible to use the standards. For example, the addition of an extra course of bricks in a wall section detail would mean that half the drawing would have to be erased and redrawn. The computer equivalent is to adjust the vertical position of the upper position of the wall, which can be done with a single transformation, and insert a segment representing the section of the extra brick. Such techniques make the use of standard details much more practicable.

The techniques of copying in different contexts, division and overlaying of information, scale changing and fast editing provide the major increases in draughting speed, but there are many small facilities that computers can provide which, while minor in themselves, will cumulatively increase draughting speed considerably. For instance, putting text onto drawings by typing is much quicker than stencilling, even for a poor typist. Lines of different styles can be drawn in a single action by selecting the appropriate option, just as a draughtsman selects a pen to give a certain thickness of line. Thus dashed lines to indicate hidden items or chained lines to define section lines can be drawn just as quickly as solid lines.

The computational and repetitive abilities of the machine can be used to perform many draughting functions directly. For example, planning grids that might otherwise take a full day of tedious work to draw can be produced in minutes. Setting out a pattern of columns, lighting fittings or other regular arrays can also be automated. Most draughting systems allow semi-automatic dimensioning of drawings; the user indicates the end points of the dimension and the machine draws the dimension line and witness lines and inserts the correct figure for its length. Various styles of hatching can be generated over the required areas of the drawing. Setting out by geometric means is far easier and less error-prone than by hand.

It must be accepted that despite all these advantages producing a single drawing on a computer system is not always faster than the manual equivalent. This is mainly because the absolute accuracy necessary on the computer in some cases requires the computer user to make more effort, when a human draughtsman could make a quick and rather rough drawing that is still perfectly acceptable. The advantages of using the computer come where the drawing can use many repetitions of a single segment, or, on the frequent occasions where there is not a great deal of repetition, at a later stage in the design process when the drawing requires modification or when larger scale plans are to be produced. It is therefore sometimes more efficient for an office to combine the use of manual and computer techniques.

Draughting systems can be divided into those that work on two dimensions, those that work in '2½' dimensions and those that work in three dimensions. Two-dimensional systems are usually the easiest conceptually and in use, and are equivalent to manual draughting in that the user arranges lines and text on a flat surface. Although very easy to use, such programs are not building any sort of model of the building in the computer's memory. For instance, a wall may be drawn as two parallel lines, but the computer is only aware of them as separate lines and not as representing a three-dimensional solid. This in turn means that certain results are unobtainable. Most importantly, it is not possible to generate views other than the one drawn, whereas a three-dimensional system might be assembled in plan, and it will then be possible to generate elevations and sections as required.

In use, three-dimensional systems can be slow and cumbersome. They require the user to define the segments as three-dimensional entities and then provide facilities to assemble the segments on screen. This definition process can take up a lot of time; a handbasin, for example, is a complex shape and could involve a lot of work to define even crudely. Three dimensional working is also especially

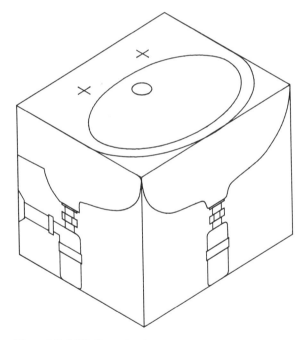

Figure 6.8 A 2½-dimensional segment

demanding in computer time as the machine has much more data to handle and has to manipulate it in a more complex fashion. For these reasons, a simplified form of three-dimensional working, called '2½-dimensional' has acquired some popularity in architectural applications. In this system, each segment is regarded as a rectilinear box, on each face of which is projected a two-dimensional view of the object represented. *Figure 6.8* shows how a handbasin might be defined in this system.

This approach has advantages and disadvantages. The advantages are that it is much quicker to define elements in this fashion than as a true solid, and also that much less computer time is required, as the machine only has to process a fraction of the information needed for three-dimensional working. The main disadvantage is that the system cannot give correct results if more than one view of a segment is visible at once on any drawing. This will occur when a segment is placed in any orientation other than parallel to the three main axes; that is, in a non-orthogonal position. Some 2½-dimensional systems go as far as completely to forbid non-orthogonal positioning, while others will accept it, but must inevitably give elevations that are incorrect in some way. In practice, this need not be a major problem in architectural applications as buildings are usually made up of surfaces at right-angles to each other, and the positioning of equipment and fittings tends to follow the orientation of the surfaces.

The 2½-dimensional systems have lost popularity in recent years. This is partly because cheaper processing power has made three-dimensional systems look more attractive. It is also because in order to appeal to the widest possible market, the vendors of draughting programs wish to make the programs usable in as many fields as possible, and 2½-dimensional systems are not very useful in engineering applications because of their limitations.

With a three-dimensional system, the user can obtain true elevations and sections after having assembled the segments into a model of the design. *Figure 6.9* shows an elevation produced automatically from the plan in *Figure 6.7*.

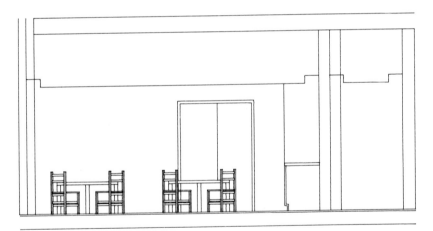

Figure 6.9 An elevation obtained from a three-dimensional system

Most three-dimensional systems are also able to operate in two-dimensional mode and this is a preferable method of working where producing views from other points is not required or inappropriate, as it is much faster.

A useful facility given by most three-dimensional systems is that perspectives or other projective views can be obtained. The 2½-dimensional systems can also produce perspectives, but since such views inevitably show several sides of many segments, correct views can only be obtained in certain circumstances. The subject of perspectives is dealt with more fully in Chapter 7.

Modelling systems

The fact that a building is being modelled inside the computer naturally leads to the thought that it should be possible to obtain other results regarding the building performance. For example, calculating the heat loss or gain should be a relatively simple matter of specifying what materials the segments are made of. It should also be possible to obtain bills of quantities, perhaps perform structural analysis and so on. Despite the attractions of this integrated approach, very few programs have gone beyond offering one or two of the potential facilities. In fact, in the UK only two fully-integrated programs have been written, neither of which could be used with general methods of building and neither of which is now supported. These are the House Design Package which was written for the Scottish Special Housing Association by the EdCAAD group at Edinburgh University[65] and the OXSYS Design package for hospitals which was written for the Oxford Regional Health Board by Applied Research of Cambridge Ltd[66]. The former program had a library of the SSHA's standard house plans built in, and the latter program assumed the use of a prefabricated building system known as the Oxford System.

The current view of the EdCAAD Group[67] is that programs of this type have difficulty being cost-effective because their huge bulk (typically several millions of lines of program) makes them expensive to implement and to maintain.

Another important reason for the failure of integrated modelling systems is that they require the user to do so much work before results can be obtained. Not only must the three-dimensional forms of the solids be defined, but the assembly must be complete with no missing or even ill-defined parts. This state of affairs does not exist until very late in the design process, by which time it is too late for most results such as for, say, environmental or cost performance, to be relevant.

Non-interactive systems

A few years ago, there were several non-interactive systems on offer. With such systems, the user does not see the drawing appear on screen as it is defined, but rather types in a set of instructions to the computer which are carried out without human intervention and the drawing produced for consideration. For example, the MEDALS system written by the Computer-Aided Design Centre in Cambridge (now the CADCentre Ltd) made it possible to perform simple two-dimensional draughting[68]. The most elementary commands were 'LINBY2' followed by two numbers, to draw a line relative to the current pen position, and 'MOVBY2' followed by two numbers, to move the pen relative to the current pen position. Thus to draw the cross illustrated in *Figure 6.10*, the user types:-

LINBY2	100	0
MOVBY2	−50	50
LINBY2	0	−100

Other commands were provided to draw circles, to define and assemble segments, and to perform most other functions.

A popular non-interactive system throughout the 1970s was the early version of the CARBS system[69]. This was a 2½-dimensional non-interactive system first used as an architectural aid by a Clwyd County Council in Wales in 1973 and later taken up by a number of other local authorities. CARBS was orientated towards building

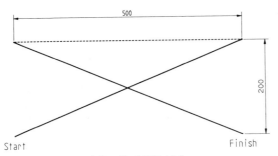

Figure 6.10 A cross defined by MEDALS

definition and had facilities for defining walls, doors, windows etc. The program automatically formed the correct junctions where walls met and could distinguish between walls of differing construction. This allowed buildings to be modelled at a more general level than that possible with simple two-dimensional systems. The user again typed simple commands, but rather than Cartesian coordinates could use the letters N, S, E and W to move along the 'compass' directions north, south, east and west. For example, to define the plan shape in *Figure 6.11* the user could type:-

    START AT A,   E 1500, N 1000, E 1000, S 1000,
                    E 1500, N 2000, W 4000, S 2000,
    START AT B,   S 1000-2T,
    START AT C,   N 1000-2T,
    FINISH

Note the use of the constant 'T' to represent the wall thickness, thus saving calculation. The later insertion of doors and windows would automatically punch the holes in the walls and form the correct junctions.

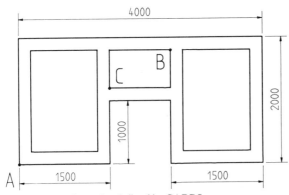

Figure 6.11 A floor plan defined by CARBS

A major advantage of the CARBS system was that detailed schedules could be produced. Schedules of materials could be output giving, say, the volume of concrete used, the area of glazing, the length of skirting, or the number of doors. The program was aware of the nature of materials and could adjust the form of output to suit. When used with traditional methods of construction CARBS could produce about 60% of the superstructure bills of quantities without human aid.

CARBS is no longer in regular use in this form. The experience of the programming team was used to develop an advanced interactive system called ACROPOLIS which is widely marketed.

A great advantage of non-interactive draughting is of course its saving in computer time and the cost of graphics screens and joysticks (once expensive items of equipment). However, non-interactive draughting is very difficult. Despite the gains from overlaying, the elimination of repetitive drawing, and so on, some tests I have made show that non-interactive draughting is about one and a half times slower than the manual equivalent[70]. Because of this, the cost-effectiveness of such systems depended on their being able to produce supplementary results such as bills of quantities.

Non-interactive draughting also suffers from the extreme tedium of preparing the data. Most people find that spending days measuring dimensions and writing out data forms is unbearably boring. Now that computer processing can be obtained cheaply and raster screens are widely available, the disadvantages of non-interactive working has resulted in its almost complete disappearance from the current scene.

Computer limitations

Despite the vastly improved performance of microcomputers over the years, the high demand that interactive graphics puts upon the machine means that they are unacceptably slow in many circumstances. A drawing will typically occupy many kilobytes, and in some circumstances megabytes, of the storage system. Simply reading this volume of information from the disc takes minutes of time on even a fast system. It is not then possible to send this information direct to the screen, because the drawing is made up of segments and two repetitions of a segment will cause lines to appear at completely different points on screen and perhaps at different angles and scales as well. Thus each line on the screen must undergo a 'transformation' process and where thousands of lines are involved this is again a slow process.

Microcomputer systems typically use floppy discs which are about one-tenth of the speed of hard discs. Their processing speed is also about one-tenth that of a microcomputer. This speed limitation is most obvious when moving about the drawing. Zooming in or out or displaying different parts of the drawing on screen requires checking all the data to find the segments that are to appear on screen, and then the transformation and drawing of every line in those segments. As this moving about is a common requirement in practice, a drawing of any size will involve many frustrating pauses in work. A recent conference on low-cost computer-aided design organised by the Royal Institute of British Architects[71, 72] showed that some architectural users are buying micro-based systems in order to obtain experience of the techniques cheaply before they purchase a large system. Two of the speakers complained of drawing times often being in the region of 5-10 minutes.

Because of the limited memory of microcomputers, it is also common for the programs to lack features taken for granted in larger systems. In some cases a suite of programs is provided to perform different tasks, but obviously this approach is more clumsy than using a single large integrated program.

There is a number of draughting programs, on both small and large computers, that hold the drawing being worked on in the computer's memory or often in memory built into the screen. The advantage of this is that many operations are much faster as the disc storage system is only accessed once, at the start of the session, rather than continually. A difficulty arises of course when the drawing is too large to fit into this memory. Some large systems can support memory of up to several megabytes and this will almost always be sufficient for the sort of drawings an architect will produce, although the cost of this amount of memory is high.

A microcomputer draughting program that holds the drawing in memory, as most do, will impose several restrictions on the complexity of the drawing. A precise limit in terms of vectors cannot be set as the repetition of a segment may add many lines while using very little memory. However, because of the high drawing times with microcomputers, the user will in any case tend to keep drawings small.

All draughting systems define a drawing in terms of full-size dimensions, using units chosen by the user. In architectural working, millimetres will be the usual choice, so a 900 mm door, say, will be stored using that number and will only be scaled down as required when plotted on paper. Systems vary, however, as to whether only whole numbers (integers) can be stored or whether the numbers may have a fractional part. Whole numbers take up only half the storage space and are faster in computation. For this reason smaller systems often restrict dimensions in this way. The problem is that whole numbers are restricted in size, usually to between plus or minus 32,768, and therefore this represents the maximum size of the thing being drawn in the units chosen. If a building site were being drawn with the units in millimetres, the site could not be more than 65 metres across. The units chosen could of course be changed to centimetres, but then the accuracy of any dimension would only be to the nearest centimetre.

This restriction can be circumvented by splitting drawings into more than one part or by careful choice of units. A system that can accept fractional units is more flexible in use, but requires more computer power to drive it. Again, because of their limitations, many microcomputer programs work only in whole numbers.

In a few years, as microcomputers become faster and larger, their limitations will be less obvious, although fast disc systems are unlikely to drop sharply in price as they are high-speed, high-precision mechanical devices. Nevertheless, the current state of affairs is that microcomputers are most suitable for less complex drawings and the larger office is better advised to install one of the more powerful minicomputer based systems.

Studies have been made of the productivity of various draughting systems[73, 74], although inevitably these quickly go out of date.

Working methods

When working with an interactive draughting program the user will be provided with a screen (or sometimes two screens) to view the drawing. On the screen will be a cursor with which the user can indicate different points. The cursor will be under the control of a positioning device which in the small systems will typically be a joystick or a mouse and on more expensive systems will most often be a digitising tablet with puck. The user will also have a keyboard for typing in commands or other text.

Screens have the restriction that they are not very large. The largest screens have a viewing area that is roughly of A3 size, and a typical microcomputer screen will be nearer A4, about one-sixteenth the size of an architect's drawing board. Because of this a drawing often cannot be worked upon as a whole, because it will not be readable. The user must zoom in and view a small part of the drawing which he will select as the area to work on. This is known as 'windowing'. A very common sequence of actions on a computer-aided draughting system is to zoom out to overview the drawing, then zoom in on the area that is to be worked on. The novice user often asks what scale the drawing is on screen: the answer is that the scale to which it happens to be drawn at at any particular moment is irrelevant. The user works at whatever scale gives the best compromise between readability and having as much of the working area on screen as possible. The drawing is always stored in full-size dimensions and can be plotted on paper at any standard scale.

Zooming and windowing are not normally done continuously, but rather in jumps. For example, the user may ask for the viewing scale to be doubled so as to zoom in. A smooth zoom, or the ability to 'pan' smoothly across the drawing would be a convenient method of working, but unfortunately can only be done on a restricted basis, if at all. This is because of computer limitations. Zooming or panning require the transformation of every line and the elimination or partial elimination of lines off the viewing area. For a large drawing, many thousands of calculations will be necessary and to give the impression of continuous motion would have to be performed at least sixteen times every second. This amount of processing power is simply too expensive for ordinary use.

Most systems therefore do not give the capability of smooth motion, and where this is possible restrictions are imposed. At least one system, by storing the drawing in the computer's memory and by only allowing whole number dimensions, has been able to perform the required computations fast enough, provided that the drawing does not exceed a few hundred lines. Above that limit, serious flicker begins to be apparent. A more general capability is provided by some systems which use advanced screens that have pan and zoom features built in. Screens of the refresh type store the vectors within themselves and it is not difficult to build in transformation circuitry. Some modern raster screens also have extra memory to store vectors and circuitry to transform them. However, as the vectors cannot be stored exactly, distortion of the image results if it is expanded more than a limited number of times. Such screens are also of course very expensive.

Having put the area of drawing to be worked on onto the screen, the user must ensure that the correct parameters for current line appearance and text appearance and size are set. Once a certain line appearance is set, all subsequent lines drawn will have that appearance. It is therefore conceptually similar to picking up a certain pen, except that not only line thickness and colour can be changed, but also the style. It is important for reasons of efficiency to distinguish between line styles that can be drawn by the hardware of the screen itself, and line styles that are drawn by sending many small vectors. Most screens designed for graphics work can automatically draw lines in a few different styles; for example, solid, dotted, dashed and chained lines. They can also draw any of these at several different brightness and in a variety of colours. However, the user may want some other style. In this case the computer must calculate and send the many components of those lines by software. This is much slower and if such lines are used widely will slow drawing time considerably. The simple raster screens usually fitted to microcomputers often support only solid-line styles.

A rather similar situation exists with text. Most screens have a hardware-generated text style, which is rather crude but readable. However, in almost all cases it is only possible to change the size of that text and not to rotate it from the horizontal. Therefore each character must be generated by software as many short vectors, and again this can greatly slow drawing time. Elaborate styles are usually so slow, even with quite powerful systems, that they will only be used on the occasional presentation drawing.

To draw a line, the user issues the appropriate command and indicates the endpoints with the cursor. With most systems, the line appears after the second point has been hit, but some programs use the 'elastic band' method. The user first locates the starting point, then as the cursor is moved away a line will appear stretched between the cursor and the starting points. Any movement of the cursor will completely retrace the line in a new position. *Figure 6.12* shows the effect diagrammatically of moving a lightpen from one point to another.

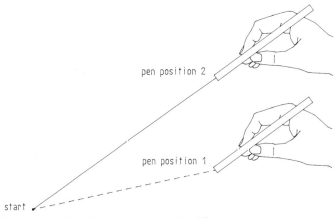

Figure 6.12 The definition of an elastic-band line

A manual draughtsman uses a tee-square and set-square to orientate lines. A computer system has no such aids, but the facility is usually provided by having special commands to draw truly horizontal or truly vertical lines. On a computer system it is easy to skew the coordinate axes so that 'horizontal' and 'vertical' can be defined to be at any angle.

A problem with computer draughting is that lines must be positioned with complete precision. This is not true of a manual system where it is only necessary to draw as accurately as the eye can reasonably see at that scale. With a computer system this is not acceptable because one of the advantages is that drawings can be blown up to almost any scale, when inaccuracies will become very apparent. Automatic dimensioning also depends on lines being precisely positioned. A facility must therefore be provided to adjust the position of line endpoints as indicated by the cursor.

The fastest method of adjusting lines is to allow the user to define a grid across the drawing, which may or may not be visible. All endpoints can then be required to be on a grid intersection. The grid spacing and its starting point can be adjusted by typing in the appropriate values, so allowing some flexibility. This system is quick and easy to implement in a program as only a very simple calculation is necessary to find the nearest grid intersection. Unfortunately, it is not very easy to work with. Many points will lie off the grid, unless it is so fine as to be difficult to use, and then either the entire grid must be repositioned or the Cartesian coordinates typed in, which is slow and tedious.

A much better solution, which is found on all serious systems, is to have a 'gravity field' facility whereby a cursor position can be automatically adjusted to the nearest endpoint of an existing line or to the nearest intersection of two existing lines. This is not an efficient solution from the computer's point of view, because when the user specifies that gravity field adjustment is to take place, every line in the drawing must be checked. It is, however, a very easy system to work with; the drawing can be built up around existing lines in a natural manner. Some draughting programs extend the gravity field concept so that the user is able to fix on to the centre of circles, the midpoint of a chosen line and so on.

Drawing arcs and circles can be specified in many different ways and in different situations it may be most convenient to specify different combinations of radius, centre, points on the circumference and tangent lines. Some systems allow any mathematically valid combination of parameters while others allow only the specification of the centre point and a typed radius value. Even if only the latter method of specification is allowed, circles can still be constructed using basic geometric techniques, but this does of course take longer and require greater skills from the user.

The basic technique of drawing lines, arcs and text can be used in the creation of segments. Different systems have different methods of doing this. Some require a segment to be produced away from the drawing, then invoked as necessary and some allow them to be defined in context, which is often more convenient. Some programs permit an hierarchical structure, where a segment can contain other segments, as when for example, a handbasin segment will contain a pair of tap segments. Others do not allow this. It is usually necessary to give a segment a unique name at the time of definition by which it will later be invoked; the permitted complexity of such names varies from program to program.

To call up a segment the user issues the appropriate command then types in the segment name; or if a copy of the segment is visible on screen, touches one of its lines with the cursor. The segment must then be positioned in the appropriate place. It must often be rotated as well, and occasionally scaled to a different size. Positioning may be done just by indicating a point on the drawing; the segment will then appear with its nominal start point, or 'origin' over that point. Some systems, however, allow the segment to be 'dragged' into position; as the user moves the cursor, the segment moves with it and so can be positioned visually. This is a convenient method of working as the user can see at once the appearance of the segment and where its origin is; mistakes are easily made if positioning is done 'blind'. Most modern screens are capable of supporting dragging, but when the segment becomes large the time needed to erase it continuously and redraw it in a new position to give the impression of movement will result in very slow movement or serious flicker. Screens that store the actual vectors within themselves, such as all refresh screens and some advanced raster screens, place few restrictions on dragging. The final positioning of the segment is subject to the same problems of accuracy as with line drawing. It is again necessary to have some sort of gravity field capability to ensure absolutely correct alignment.

Rotation will most often be in steps of ninety degrees in architectural work, but all systems allow arbitrary rotations. The angle may be typed in, or may be parallel to the current axis system, which can itself be rotated. Scaling of segments is relatively rare: normally the user will choose to define the various different standard sizes of an element and use those. For example, the user could in theory scale the representation of a door to any required size, but in practice it is better to have a library of, say, four door sizes and use those. This imposes rationalisation, saves the effort of correct scaling and also allows a list of door segments for scheduling purposes to be obtained more conveniently. The main exception to this is with trees or bushes on landscaping drawings. The arbitrary variation in height and spread is most easily done by scaling a single segment.

By use of techniques such as the above, a drawing may be constructed. Almost always it will be necessary to alter it in some way at a later stage, and often the alteration will be drastic. Deletion of individual lines or individual segments is very easy. The correct command is issued and the line, or with a segment one of the lines, is touched with the cursor. Repositioning of a segment is also a very simple and very powerful facility. The appropriate command is issued, then a line on the segment touched to identify it. The segment can then be dragged to its new location or otherwise transformed. Deletion and repositioning are the most used elementary editing commands, but a variety of others will be provided, according to the sophistication of the program.

At a higher level, most systems provide a 'global editing' facility in which many lines or segments will be affected. Obviously such commands, if incorrectly issued, can destroy an entire drawing. It is therefore important that some sort of 'clearing' facility be available, whereby all the changes made are reversed. Ideally, the situation will revert to where it was before that particular command was issued, otherwise to where it was before that particular interactive session commenced.

The first step in global editing is to identify the area to be affected. Some systems affect all lines on the screen at that time; many others allow the user to define a rectangular area on the screen within which editing will take place; while the most advanced systems allow the user to define an area of any shape on screen. This latter ability is relatively rare, but is the most convenient as it is often necessary to work on odd-shaped areas, such as a group of rooms.

A range of global editing commands will be available. The simplest will be to erase all lines within the defined area in the same way as a draughtsman would use an erasing shield in manual work. Other basic commands will be to move all lines and segments within the area; thus if a room changes in size, it might be possible to adjust the position of one wall and all the equipment along it with a single command. Copying everything within the area is also often useful, as when two rooms are identical. Copying a mirror image is a variation on this and is again often useful as buildings often contain symmetric sections. Other commands may be available to, for instance, change the appearance of all lines or text.

At an even higher level of sophistication, some systems allow selective changes to be made within the defined area by providing the facility to declare that only certain types of segment or certain types of lines or text will be affected.

When working with three-dimensional systems, certain concepts have to be extended. The most basic of these is the concept of windowing so as to obtain the required view. On a two-dimensional system the only thing to be defined is the rectangular area of the drawing to be displayed on screen. In three dimensions, the direction of view can change, so the same drawing can be viewed in plan, in an elevation or even in an arbitrary direction such as perpendicular to a pitched roof.

When the direction of view has been established, a rectangular area must be defined to get more or less of the drawing onto screen, in the same way as in two dimensions. Finally, it is often necessary to take a 'slice' out of the depth of the drawing; for instance if working in plan on a multi-storey building, a slice corresponding to a particular floor will typically be required. All three-dimensional systems provide methods of windowing, but many are difficult to use in practice.

Given the drawing on screen most actions require the user to specify the third coordinate as well as the two that can be obtained with the positional indicator. Thus in placing a segment representing a shelf, its height relative to ground level must be specified in order for it to be added to the model of the building. It may also be necessary to rotate a segment about one or all of the three axes; if the axis is not along the direction of view this cannot be done visually. The third coordinate can of course be typed in, but this is slow and error-prone. A method very often used is that the user can define a 'base level' at any point along the direction of view; so if viewing in plan, the base level will typically correspond with the floor level of the area being worked on. Once the base level has been set, all segments added to the drawing will use it as the third coordinate of the segment origin. Many systems take this a little further by allowing the user when defining a segment to specify the height it will normally have above the base level and this will be added on automatically when positioning. This method works quite well, especially in architectural applications where most components such as walls, doors and furniture will stand on the floor and many others, such as windows, will be at a fixed height above the floor.

An alternative method of working which is more flexible than the base level concept, but is more expensive to implement, is to provide the user with two views from different directions. These two views may be in different parts of the same screen, but this is inevitably cramped, so usually two screens are provided. Typically, the user will have a plan view on one screen and a side elevation on the other. When a segment is invoked, it will appear on both screens so all coordinates can be fixed visually. The second view direction must be perpendicular to the first view direction, but can rotate about the axis formed by the latter. For engineering work, this method of working is very convenient as the components in an engine, say, do not share levels to any extent and also often need to be rotated to odd angles.

Drawing transfer

As more and more organisations use computer-aided draughting, so the advantages of transferring drawings between systems becomes more obvious. At a very early stage in the design process, a map of the site will be needed. Manual methods involve obtaining the printed map, enlarging it photographically to the required scale, then tracing off the required information. However, many maps are now obtainable in digital form; the UK Ordnance Survey currently have about 18,000 of their 1:1250 and 1:2500 maps available in this form, including most of the London area, and are adding to them at the rate of 2,000 a year.

At a rather later stage, the architect will commission a survey company to measure the site very accurately, including the exact position of its boundaries, its ground levels and any existing buildings. Almost all large survey companies now use electronic methods to gather their data; it is now possible to obtain theodolites and tacheometers that will record digitally the bearing and distance of the point they are aimed at. This information can then be passed to a computer and an accurate base survey plotted out[75]. Sometimes the survey is then carried forward manually, but many surveyors have invested in computer-aided draughting systems and finish the drawing on the screen. Again, when working manually, the architect must trace off the required information on to a new drawing sheet, but potentially the whole survey can be transferred with little effort and no error. It is then of course possible to check dimensions with the assurance that the dimensions will be as accurate as the surveyor can measure them.

'Downstream' of the architectural process, drawings may have to be exchanged with the structural engineering consultant. Almost all engineers use computers for their calculations and it has been a natural extension of this to use them for draughting as well. Once again if, say, the structural grid and column and wall layouts can be automatically transferred, there are obvious advantages.

The mechanical and electrical services consultants tend to make less use of computer techniques because of the nature of their work. However, the advantages of computer-aided draughting are becoming apparent and a number of firms have invested in a system. In this case, the advantages of automatic transfer are almost entirely of benefit to the consultant rather than the architect, but it is often possible to negotiate a fee for the service and of course the savings in time will benefit the project generally.

There are hundreds of draughting systems on offer at present, and many of them specialise in certain fields such as architecture or surveying. For this reason, it will rarely be possible to transfer drawings in a straightforward way. Unfortunately, there is no simple solution to this problem at present and a number of methods is in use to effect transfer. Essentially all transfer systems reduce to a method of converting a drawing into alphanumeric characters. The transfer of alphanumeric data is well established, then at the other end the data must be converted back into a drawing on the new system with as little loss of information as possible.

Some organisations have laid down and published their own interface format and expect users of their services to conform with it. The more popular draughting systems have so much of the market that in many cases the smaller systems have been forced to provide output options that conform to the command set of the larger systems. There is a natural reluctance to do this, but market pressures often force the issue.

Many firms with some programming expertise have written programs that take the plotter output of a system, which is normally in alphanumeric form, and convert it back into a set of commands that will drive a particular draughting program. This is the only way that communication can be established if the producers of a draughting system cannot or will not provide an alternative. The problems are that plotter output is not absolutely precise, being typically to the nearest tenth of a millimetre on the paper, which gives an inaccuracy of, say, ten or twenty millimetres when converted back to a dimension. It is also probable that the segmentation structure will be lost, as a plotter has no need for such information. Of course, if the program outputting the data is a three-dimensional one, the third dimension cannot be recovered by these means.

The solution that will finally resolve the problem is the use of an agreed standard. Several have been proposed, but the leading one is the Initial Graphics Exchange Specification (IGES) which is now a draft international standard[76]. Because it has to provide methods to handle the features found on the widest possible range of programs, it is very large and complex and this has hindered its introduction. However, the need for some form of free transfer is so great that most large systems have provided or are in the process of providing IGES format input and output options.

The AutoCAD system

The AutoCAD program, written by Autodesk Inc., is the most popular computer-aided draughting system in the world, with over 40,000 workstations currently using the package[77]. It owes its popularity to the facts that it is very comprehensive and supports many advanced features not found on rival systems and yet is cheap even by microcomputer standards. The basic software can be bought for the equivalent of about five weeks' salary of a typical draughtsman, and a complete system including both hardware and software for about a year's salary. The system is widely marketed, including by IBM[78].

The program runs on an IBM PC and compatible microcomputers. Because of the relative slowness of such machines, regular users will upgrade the machine to have 640 kilobytes of memory and a hard disc, rather than the floppy discs normally used. For use in architectural applications, where drawings are often large, it is also advisable to fit an extra piece of circuitry called a maths co-processor, which will considerably improve performance.

A high-resolution screen makes working with detailed drawings much easier and AutoCAD will support screens having up to 1,024 by 1,024 dots resolution and 256 colours simultaneously. When working, a narrow band at the bottom of the screen will normally be used as a 'dialogue area' where typed commands and the program's replies will be displayed. Various pointing devices are supported, but the architectural user will find a digitising tablet is the fastest and the most flexible of these.

The program operates on a single drawing at any one time. The drawing may consist of any number of overlays, or 'layers' as they are called, each of which is given a unique name by the user. The restriction is imposed that a given overlay can only support a single style of line in a single colour. Thus one overlay might contain solid lines in a certain colour and another dotted lines in the same colour. The user therefore typically uses a group of AutoCAD overlays to make up a single logical drawing overlay, such as service outlets. Individual overlays can be displayed on screen or not as required.

The program stores the drawing coordinates as floating point numbers and so is accurate to six significant figures. This gives more than adequate accuracy for building purposes.

On entering the program, the Main Menu is displayed. The most important options on this are to change an existing drawing and to begin a new drawing. In both cases the name of the drawing must be typed in. The drawing, or a clear screen for a new drawing, is then displayed.

The user can now select commands as required from the menu on the tablet. Various optional software extensions are available for AutoCAD, and this description assumes the full specification is used, when about 300 commands are available to perform various functions. At the most basic level, straight lines can be drawn. The command 'LINE' is selected and the user then indicates the start of the line. As the cursor is moved, a rubber-band line will be generated and the end point of the line can be fixed. A further line can then be drawn from that point, and so on until the return key is pressed.

Various forms of 'snapping' to exact points are provided. The simplest is to force snapping to grid intersections. The spacing, origin and rotation of the grid can be varied as required. More powerfully, a facility called 'object snap' will perform a search within an area around the cursor for a specified feature. The object snap mode is turned on by the 'OSNAP' command and the user can specify how searching will be performed. Ten forms of searching are available, but the most useful are likely to be snapping to the nearest point on a line or arc; the nearest end point of a line or arc; the intersection of two vectors; the centre of an arc and the tangent to an arc from the last point specified. Snapping is available for all commands that require a position to be indicated, thus if the user is drawing straight lines, a letter can be typed before indicating a point on screen to specify which type of search should be used to refine that point.

Other elementary draughting commands are those to draw circles and arcs. To draw a circle the user must first issue the 'CIRCLE' command, and then has several methods available to define it: the centre point plus a typed radius value or diameter value; three points on the circumference or two points across a diameter. Arcs are drawn in a similar fashion using the 'ARC' command, but other possible parameters are the included angle and the chord length.

Various other elementary draughting commands are available. The most important of these is the command to place text on the drawing. Another, which is very useful but which is not found in most draughting programs, is a command to draw a line of arbitrary thickness. The command 'TRACE' allows the user to type any thickness value, then to draw a chain of lines in the same fashion as for the LINE command. Other commands allow the user to place dots ('points') on the drawing and to create solid-filled regions.

At a higher level of organisation, the user can create segments, which are called 'blocks' by AutoCAD. A segment is created by assembling graphic elements on screen. Essentially the user can either point to each element that is to be included in the segment, or can define a rectangular area on screen, when all elements that lie completely within the area will be included. The specific order of operations is that the command 'BLOCK' is issued. A unique name for the segment must then be typed in and also a position must be given which will be considered as the origin of the segment. The user can then go on to touch each graphic element in turn, terminating the sequence by pressing carriage return, or can type 'W' followed by the lower left and upper right corners of a rectangular area.

The graphic elements specified may be elementary entities, such as lines or text, or may themselves be segments. By these means segments may be nested. A limit of ten levels is placed on nesting.

Once a segment has been defined, it can be repeated any number of times on the drawing by the command 'INSERT' followed by the segment name. On repetition, the position, scale and rotation of the segment can be defined.

A range of powerful commands is provided to make changes to an existing drawing. Most of these have a similar form to the segment creation command in that the user can touch a series of items to be affected by the command or can define a rectangular box on screen to enclose the items. The most basic modification command is 'ERASE' which removes elements from the drawing. Other useful commands are 'MOVE' which moves the elements to another position; 'ROTATE', 'SCALE' and 'MIRROR' which alters their appearance, and

'COPY' which makes a copy of the elements in another position. The use of these simple and consistent methods of editing means that drawing modification is both fast and user-friendly.

One difficulty with editing is that the commands operate at the highest level of organisation. That is to say, if ERASE, for instance, is in operation and a single line on a segment is touched, the whole segment will disappear. For this reason a command is provided to disassemble a segment into its component parts. The method is to use the INSERT command already described, but to precede the segment name by an asterisk. This will call up and place the elements of the segment, but these will be individual items that can be edited separately. Thus in use, to alter a single line in a segment the user would delete the existing version and call up another copy in the same position in its disassembled form. If required, the new version can be reassembled into a new segment with the BLOCK command.

When working on a drawing, it is necessary to be able to display specific portions of it on screen and to move from portion to portion. Two principal commands are provided for this: 'ZOOM' which changes the scale of the drawing on screen and 'PAN' which maintains the scale but moves about the drawing. It is also possible to give a name to views of portions of the drawing. Thus if a certain area of the picture is often isolated on screen, it can be given a name and the command 'VIEW' followed by that name will display the area.

The problem with moving about a drawing is that redraw times can be considerable. In order to ease this problem, AutoCAD provides two commands to speed up redrawing at the expense of losing some detail. These commands are 'FILL' and 'QTEXT' (for 'quick text'). If FILL is switched off, solid-filled areas are shown in outline only, and if QTEXT is in operation, lines of text are indicated by a pair of parallel lines. The careful use of overlays is also a very powerful means of speeding redrawing. The 'LAYER' command can be used to suppress the display of certain overlays and later to show them again. Thus if a drawing has been split into logical parts, much of it can be made temporarily invisible for speed and convenience of working.

AutoCAD is a large and professional program and contains many other features such as semi-automatic dimensioning, freehand sketching, automatic hatching and so on. One apparently minor, but in practice extremely useful, command that can be mentioned is the 'OOPS' command. If an ERASE has been wrongly issued so that elements are deleted that should not have been, this command will reverse the effects of the erase.

AutoCAD is essentially a two-dimensional draughting package, but recent versions incorporate some elementary three-dimensional facilities. The command 'ELEV' will extrude a plan shape into a prism so enabling simple block models to be constructed. It is then possible to obtain perspective views of the model from any point.

The GDS system

GDS ('Graphics Design System') is a computer-aided design package, the main feature of which is a powerful 2-dimensional draughting system. GDS includes various other facilities of use to architects, including 3-dimensional visualisation, space planning and reinforced concrete detailing. The package was originally written by Applied Research of Cambridge (ARC) Ltd. which has recently been acquired by McDonnell Douglas Information Systems.

The origins of GDS go back to 1971 and it was launched in its present form in 1979. It is a large and powerful program and most installations have been on fast minicomputers; originally the Prime series and more recently also the DEC VAX series. It is also available on MicroVAX microcomputers for the smaller office. About 350 copies of the program have been sold worldwide.

A typical configuration of a minicomputer, two workstations and a plotter will cost the equivalent of 6-8 years of a draughtsman's salary. This is not a small sum, but the productivity increase will be correspondingly great. One study has found that GDS is typically five times faster than manual draughting[26].

GDS is capable of driving a variety of screens, but the most popular arrangement at present is to have two screens per workstation, one to show the drawing and one to show the printed dialogue between the user and the computer. The dialogue screen will be a low-cost visual display unit and the graphic screen will most often be a high-quality colour raster screen with a resolution of perhaps 1024 by 768 dots.

A graphics database in GDS is called a 'Drawing File' and can hold a number of separate drawings. Most other systems can only hold a single drawing in one file, but the GDS approach simplifies administration as a batch of associated drawings, such as a set of floor plans for a single building, can be kept together. Drawings can be overlaid freely to make up a complete image. However, each overlay remains a logically separate entity and segments (or 'objects' as GDS calls them) cannot exist partly on one drawing and partly on another. Each drawing can of course use a full range of colours and line styles. When several drawings are overlaid, one must be declared to be 'editable', that is, capable of modification, while the others form a background.

When setting up a Drawing File for the first time, the user must specify if the drawings within it are to be stored to 'single precision' or 'double precision'. Single precision provides six significant figures of accuracy and is adequate for most architectural purposes. Double precision gives twelve significant figures of accuracy and is most used for mapping. It may be noted that this degree of accuracy is capable of recording an error of one foot between the earth and the sun!

The program is divided into a number of subsections, each of which supports a certain type of operation. The most used subsection is called 'XGRAPHIC' and provides commands to add elements to a drawing and to modify a drawing. The subsection XGDS provides administrative commands; the subsections XBLOCK and XOBJECT provide global editing facilities and so on. The subsection name must be given to the program before its commands are available.

In GDS, the user indicates a point on screen by moving the cursor to that position then pressing a letter on the keyboard. According to the letter given, a different type of search will take place to refine the point. For example, the letter 'D' specifies that no searching is to take place; the letter 'L' that the point is to be adjusted onto the nearest line; the letter 'P' that the point is to be adjusted onto the nearest line end point or intersection of two lines and so on. A total of eleven different letters (called 'hitcodes') are available to carry out various types of searching. When mastered, this single-keystroke method of working is both fast and accurate.

In the XGRAPHIC subsection, one of the most elementary commands is 'LINE' which provides the facility of drawing straight lines. The user indicates a starting point with the cursor and a hitcode, then goes on to indicate subsequent end points. Straight lines will appear between each pair of end points and the chain of lines can be terminated by pressing carriage return.

The designers of GDS aimed to make it a fully comprehensive system: the user should not have to escape into geometric constructions or other roundabout means of achieving a result when the computer could perform the calculations. This philosophy is clearly seen in another elementary command: that for drawing a circle. The user issues the command 'CIRCLE' and is then able to construct a circle by almost any conceivable means. To take one example, a circle can be defined to be tangent to three existing circles. It is virtually impossible to obtain this result accurately by geometric methods, yet few if any other draughting systems provide this facility.

A very wide range of other commands is available to perform graphic creation and manipulation at a basic level. In fact, the XGRAPHIC subsection provides 86 different commands.

At a higher level of organisation, GDS imposes a rigorous system for segmentation. All elementary graphics, such as lines, arcs and text, must be placed in one segment or another. It is not possible, as with many systems, to only use segments when it is known that a section of the drawing will be repeated in another context. Another restriction is that segments may not contain other segments. As will be seen, these restrictions enable future modifications of drawings to be faster.

Once defined, a segment may be placed anywhere on a drawing and at any scale or rotation. GDS in common with other draughting systems stores coordinates at full size, but it has the further level of sophistication that a record is kept of the scale at which the drawing will normally be plotted on paper. A drawing might therefore include a central portion at 1:100 scale; a frame at 1:1; a block plan at 1:500 and a detail at 1:5. This is not possible with most other systems, which would have to distort the size of ancillary portions of the drawing.

It is in modifying a drawing that GDS attains its highest efficiency. Modification commands are grouped into threes, each operating at a different level of organisation. Taking the example of erasure, there are three commands: DELLINE, DELBLOCK and DELOBJECT. If a line is touched when DELLINE is in operation, then that line will disappear. If DELBLOCK is in operation, then the entire chain of lines of which the indicated line is a part will disappear. If DELOBJECT has been selected, then the entire segment of which that line is a part will be erased. This system is much faster than that used by most other programs, where the removal of a single line involves exploding a segment into its component parts, deleting the line, then reassembling the segment. A symmetry of naming ensures that the higher number of commands is not a difficulty to the user.

It will be clear that there is a potential problem with DELLINE and DELBLOCK in that the drawing may be left with two segments having the same name but different contents. GDS deals with this by adding a 'version number' to the segment name. This is done automatically whenever differentiation is necessary.

The separate subsystems XOBJECT and XBLOCK are provided to perform global editing. A range of commands is provided to change the drawing and also to limit changes in different ways. The limits may be in area: the user can define a polygon on screen within which changes may take place. The limits may also be on types of graphic element (text of given size or colour or lines of given style, thickness or colour) and on segment naming.

Many other facilities are provided by GDS which is a very comprehensive program indeed. Some of the most useful include automatic hatching, automatic dimensioning, a built-in BASIC interpreter, interfaces to programs written in any computer language and to a database system and the ability to define any line or text style.

Because of its comprehensive nature, GDS is not an easy program to learn. The manuals contain a total of about 700 pages detailing over 400 commands. However, once skill has been acquired it will be found to be a very fast and flexible way of working.

The Intergraph system

The Intergraph Corporation is one of the largest computer-aided design system vendors in the world. The firm was founded in 1969 and currently has well over 2000 systems installed worldwide[79].

The hardware offered by Intergraph has shown great consistency over a long period. The company has standardised on DEC computers since 1977 and at present offers a range of configurations of different power and price within the VAX range of minicomputers and the MicroVAX range of microcomputers. The workstations themselves are purpose-built and consist of two identical screens within the same casing, mounted over a large digitising tablet. The visual effect is of an integrated system, in contrast to the products offered by almost all the smaller vendors who support a wide range of different computers and screens and whose workstations therefore often give an uncoordinated impression. *Figure 6.13* shows a workstation in use.

The Intergraph system incorporates some very advanced hardware in order to achieve high performance. The workstations themselves contain a powerful microprocessor and also a considerable amount of local memory. The magnetic disc is equipped with a piece of circuitry called the 'File Processor' which filters the data when a drawing is accessed and so greatly reduces the load on the main computer. The File Processor, for example, can block the transmission of parts of the drawing which are currently off-screen.

As might be expected, systems of this sophistication are not cheap. However, it is a fact that computer-aided design and draughting systems need a great deal of power if their users are not to have to put up with constant pauses in working.

The principal suite of programs is called IGDS, for Interactive Graphics Design Software. It has the ability to create both two-dimensional and three-dimensional drawings and manipulate them freely. A single drawing will be contained within a named file, known as a 'Design File'. The drawing may be structured as up to 63 separate 'design planes'. In two-dimensional working, these can be thought of as conventional overlays. When working in three dimensions, they can be visualised as 63 exactly coincident transparent cubes.

IGDS has an unusual method of storing coordinates. They are held as 32-bit numbers, which is equivalent in storage terms to floating-point precision, but the numbers are divided into three parts: the Master Unit, the Sub Unit and fractions of the Sub Unit. Typically, the Master Unit will be metres and the Sub Unit millimetres.

When working, a puck and a tablet will be used to select commands and to indicate points on either screen. The dual-screen configuration is one of the most useful features of the Intergraph system, as a number of aspects of the same drawing can be displayed at once.

Figure 6.13 An Intergraph workstation

Most simply, one screen can show the interactive dialogue and the other screen a drawing. However, it is also possible, for example, to show the whole drawing on one screen and a small portion of it on another. The user can work on the large-scale view in comfort, then turn to the overview to move to another part of the drawing.

It is also possible to quarter a screen and show a different view of the drawing in each quadrant. This arrangement is most convenient in 3-dimensional working, where the views might typically be the front face; the top face; the right-hand side and an isometric projection. Specifying a point in 3-dimensional space can now be done by two cursor hits. The first hit drills a 'borehole' into one face, the line of which is reflected on the other views. A second hit on that line will then fix the point. This is a fast system in use and is not prone to error, unlike systems which are restricted to one view at a time and so give the user no depth cues.

The usual basic graphic elements can be added to a drawing, including straight lines, circles, arcs and text. It is also possible to sketch free curves, which are very useful for road lines, boundary lines and so on. Other basic elements include ellipses, parabolae and spirals, but these are more applicable to engineering applications than architecture.

The facility of snapping on to exact points is provided at two levels. It is possible to refine points directly by the use of 'locks'. The most powerful lock is the 'snap keypoint' lock. When this is in operation, the computer will search around the cursor for line or arc endpoints, text justification points and segment origins. Each point found is highlighted in turn and the user can accept it or go on to the next. Snapping can also be done indirectly by the use of a wide range of geometric construction commands. For example, the 'construct point at intersection' command requires two lines or arcs to be identified, then calculates the intersection and places a marker there. The marker can then be snapped to by the use of a lock.

IGDS allows parts of the drawing to be placed into segments, or 'cells' as they are called, as required. Segment definition is done by placing a boundary around an area of the drawing and declaring that the elements within it form a segment. A unique name must be given, and a short description can also be provided to aid future identification. The boundary may be of any shape: it is not restricted to a rectangular box as with many systems. The elements may themselves be segments, thus enabling segments to be nested within other segments.

When defined, segments are placed in a separate file called a 'cell library'. Users will typically build up cell libraries of standard drawing elements for use on a variety of drawings. Segments can of course be repeated at any point on a drawing and at any scale or rotation.

A range of commands is provided for modification of a drawing. They include copying of elements, moving them from place to place, rescaling, deletion and so on. These commands operate at the highest level of organisation, thus if a line is part of a segment, the whole segment is affected by the operation. If the user wishes to change a segment, therefore, it must be exploded into its component parts. This is done by the 'DROP' command which disassembles a complex element. If a segment is 'dropped' it splits into its component parts, and if a chain of lines is dropped it splits into individual vectors. These can be edited, then if necessary a new segment reassembled.

Assembly is made easier by the ability of IGDS to move elements between overlays. At any time, one overlay is 'active'; that is, only the parts of the drawing on that overlay can be referenced. If a 'MOVE' command is issued, then the active overlay changed, the element indicated will move to the new overlay while retaining its visual position. In practice, this is often used to move a segment to a spare overlay before disassembling it so that it can be modified and remade without confusion with other parts of the drawing.

IGDS provides a range of commands for area editing. An arbitrarily-shaped boundary can be drawn around the area. Then its contents as a whole can be moved, copied elsewhere, deleted and so on.

In order to simplify working, IGDS is able to make any combination of overlays visible or invisible on screen. In 3-dimensional working it is also possible to take a 'slice' through a drawing. This is done by specifying upper and lower limits along one view. In architectural working, this 'display depth' as it is called will typically correspond to a certain floor within a building. Within the display depth, the user must also define an 'active depth'. This is a plane on which points are assumed to be located unless the user specifies otherwise. Using the display depth as a working surface enables 3-dimensional points to be defined by a single positional indication rather than two.

A further means of helping presentation is the ability to dynamically alter the way in which the drawing is viewed. The drawing is stored in the memory local to the screen and the microprocessor associated with it can rescan it at high speed to give the impression of continuous movement. The user can expand or shrink the area viewed up to eight times. It is also possible to pan across the drawing and, in 3-dimensional working, dynamically rotate the view on screen.

IGDS is a comprehensive program and also contains many other commands to perform automatic dimensioning, hatching and so on. It also has links with Intergraph's database management and retrieval system, called DRMS, which allows numeric or textual attributes to be associated with graphic elements. DRMS can be used to answer queries, perform analyses, print reports and so on. Information can be passed from an entire drawing or from a restricted portion of a drawing. It is therefore possible, for example, to produce equipment schedules on a room-by-room or on a departmental basis.

Chapter 7

Visualisation

Principles

It is an interesting contradiction that although the most important aspect of architectural design is the attractive disposition and massing of three-dimensional solids, an architect develops a design almost entirely from the two-dimensional plans and elevations. Perspectives may be drawn during the sketch design phase, but these will usually be rather rough and poorly detailed. Axonometric views are also sometimes drawn. This type of projection is popular with architects as it is easy to draw; it does however, have the drawback that as only aerial viewpoints can be used, the drawings are of limited relevance unless the building is regularly seen from a helicopter. Essentially, the architect relies on his spatial sense to tell him if the building is visually acceptable.

Properly drawn perspectives or detailed models may be produced to present the building to the client and others whose spatial sense may be less well developed; but because they are expensive to provide and difficult to modify, they are nearly always produced after the design is finalised and play no part in the generation of the design.

The methods of generating a representation of a building by computer, techniques known as 'visualisation', can be used to produce views from any point and from any angle very quickly. These views can help the architect in generating a pleasing design; they can also be used in presenting the building to the client and to the other members of the design team, thus increasing involvement from an early stage.

Computer visualisation can sometimes offer results otherwise unobtainable without a great deal of effort. For instance, it can give representations from unusual viewpoints from which it would be difficult to calculate perspectives, or which would be inaccessible to a modelscope. Once the initial data has been input, generation of perspectives is fast and cheap, so that many views can be produced with little effort. The designer can therefore 'walk round' the building and see it as an observer would after completion. The more advanced visualisation programs can depict colour and texture, for example that of brick or glass. Some can also show the shadows cast at different days of the year, times of the day and at different latitudes. If a building has strong shadows this information is very valuable to the designer but would require extensive calculations to draw manually.

If data is provided for internal walls it is also possible to obtain perspectives of interiors. From these, the architect can see the effect of 'walking through' the building, a consideration that is often neglected when designing.

There are several different basic forms of computer visualisation that may be produced. The simplest is known as a 'wire-line' perspective. The effect given is as if the object were transparent, with the lines that are normally hidden being shown. A typical wire-line building perspective is shown in *Figure 7.1*.

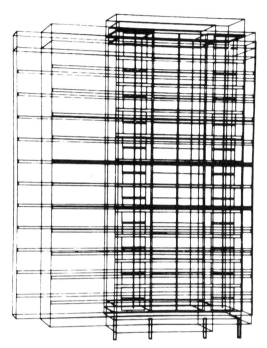

Figure 7.1 A wire-line perspective (courtesy CADCentre Ltd)

The generation of this sort of perspective is very quick by machine as a simple mathematical transformation will convert lines in space into lines on a two-dimensional plane, representing the object as seen from a particular viewpoint and in a particular direction. Relatively few calculations are involved, and there need be no limit on the complexity of the scene.

To remove the hidden lines and show the object as if it were opaque, the computer must perform many more calculations. In the worst case, every line will have to be compared with every plane surface to see if any plane hides it or a part of it. Over the years, a lot of effort and ingenuity has gone into solving the problem of hidden-line removal; it is now well understood and it is unlikely that there will be any great improvement on the methods now used[80]. However, even the best of these is comparatively slow. The equivalent hidden-line perspective of the building in *Figure 7.1* is shown in *Figure 7.2*.

Almost all hidden-line removal programs place a limit on the complexity of the scene. This is so because for efficiency all the data is normally held in the computer's memory. The number of calculations necessary is not only very high, but increases exponentially with the complexity of the scene. If data had to be repeatedly recovered from disc memory for processing, the computing times would become intolerable.

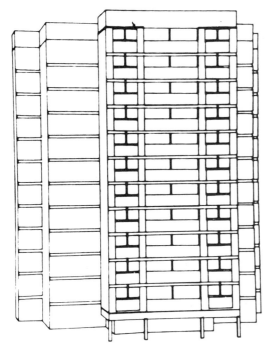

Figure 7.2 A hidden-line perspective (courtesy CADCentre Ltd)

Another form of visualisation output is that in which the surfaces are shaded or coloured, so giving the effect of a photograph rather than of a line drawing. This sort of output is called a 'hidden surface' representation and an example is shown in *Figure 7.3*.

Surprisingly, the generation of hidden-surface results is considerably faster than that of hidden-line results[81]. This is because a half-tone picture will inherently go to a raster screen and so it is not necessary to calculate more precisely than the dot size of the screen. A typical raster screen might contain a matrix of 800 by 600 dots, so in the worst case each plane must be sampled about half a million times. This still represents a heavy computational load, but not as much as is involved in calculating exact intersection points as is necessary in hidden-line processing. Further, the computation time increases only linearly with the complexity of the scene rather than exponentially.

Hidden surface views can be in monochrome or in colour. The user may be able to specify the colour of each plane and also its reflectance. Given the source of the illuminance (normally the sun) the computer can derive the appearance of the scene, where planes facing away from the light are less bright. Some advanced systems allow the user to specify a type of material and will then give the correct textured appearance to the view[82]. The most sophisticated hidden-surface programs take the process one step further by calculating the position of shadow patches and modifying the tone of the surfaces they fall on accordingly[83].

One problem with hidden-surface output is that the computer user needs a sophisticated method of driving the screen. Many microcomputers use three bits of information to drive each dot on screen and so there can only be eight different

Figure 7.3 A hidden-surface perspective (courtesy CADCentre Ltd)

colours or shades of grey on the screen at once. This inevitably gives a garish and unsubtle representation. More advanced microcomputers use eight bits per dot and so can support 256 colours or shades of grey. This is much better and will give a tolerable result; however, 256 colours may sound high, but because allowance must be made for colours muting as they move into shadows, it still does not allow for much variation. Best results require an intelligent raster screen with its own memory. Some advanced screens of this type, such as the Tektronix 4125, can hold 24 bits per dot, giving a capability of over 16 million colours. With this sort of device results can be obtained that approach those of a photograph. Such screens are of course expensive, and although they can also be used in conjunction with a computer-aided draughting system, not many offices will be able to justify one.

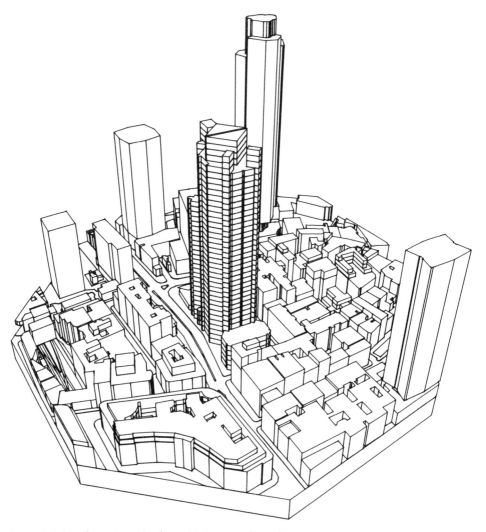

Figure 7.4 An urban perspective (assembled by Nick Gibson)

If the computer has enough capacity, a building can be shown in the context of its surroundings. This can be important on a high-density urban site where many large buildings are present. *Figure 7.4* shows a perspective of a building on an urban site.

The generation of interior perspectives is not different in principle from the production of external views. An example of an interior view is shown in *Figure 7.5*.

When looking from one room into another, it is often more helpful to see the hidden lines than to have them removed as a better idea of the circulation patterns can be gained from them. Most visualisation programs provide the option of retaining the hidden lines to give a wire-line drawing. The presence of many extra lines can be confusing, so some programs provide the useful compromise of showing the hidden lines dotted, while leaving the visible lines solid.

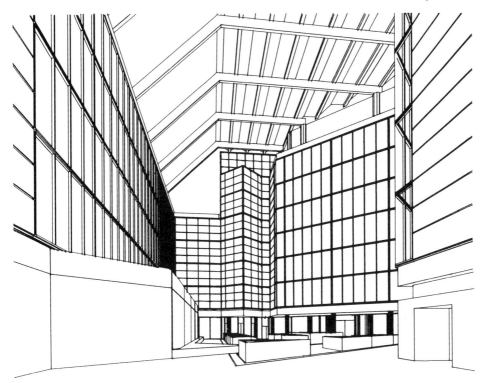

Figure 7.5 An internal perspective (assembled by Alan Minnerthey)

As has been pointed out, hidden-line removal is demanding of both memory and processing power and for these reasons problems will be experienced when using microcomputers. To take the example of the (admittedly complex) interior perspective above, it contains 6,000 separate planes and occupies about 300 kilobytes of memory. This amount of memory is only available on advanced microcomputers. To produce the result took four hours on a fast minicomputer which operates at around 0.65 million instructions per second. Even an advanced microcomputer would take three or four times as long.

Speed is not necessarily a problem in this application. A useful way of working is to produce wire-line results in order to establish the views required – these will typically require only a few minutes each – then leave the computer running overnight to generate the hidden-line or hidden-surface outputs. As perspectives are needed only irregularly, this will normally be quite practicable. If, however, a long series of views is needed, for example for a film, then speed restrictions will be a serious problem. In such cases it may be worth investing in a machine that performs hidden-surface removal by hardware. Such machines have been used for many years in aircraft simulators and for other applications, but have been far beyond the resources of any architectural firm.

Fortunately, the use of microprocessors has now brought these devices to an affordable level, although even so they cost the equivalent of several man-years of staff time. One such machine is the Tektronix 4129. This contains a microprocessor and up to 800 kilobytes of memory to store the data. It has many capabilities,

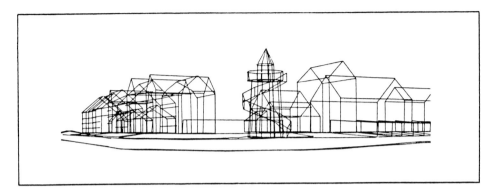

VIEW FROM ISLAND

Figure 7.6 A computer-produced perspective and its hand-finished version (courtesy David Ruffle Associates)

including hardware hidden-line and hidden-surface removal. The screen is a very high resolution raster screen containing 1,024 lines of 1,280 dots, and can display up to 1,024 different colours.

In most cases, memory capacity will be a far more important problem than speed. Allowing for the size of the visualisation program itself, most microcomputers will be unable to handle more than 500–1,000 planes, and as even a single doorway might be made up of over 100 surfaces there is little scope for detail. This means that it will be necessary to define the building in broad outline, avoiding intricacies.

One of the most serious drawbacks of computer visualisation is that the quality of the result will not be good as the manual equivalent. The lack of detail is one problem. This applies not only in the building itself but also because there will be no question of putting in trees, figures, cars and the other small items that add finish to a perspective or model. Also, every line in the output will be the same thickness, where a human draughtsman will almost automatically draw such features as doors, windows and facade details more lightly than the rest to balance the emphasis of the design.

In practice, this poor quality will not usually hinder the designer, who will be more interested in the overall impression of the building. A computer perspective may not be considered good enough for publication or formal presentation, but savings can still be made by using the computer output as a basis for manual finishing. An example of such finishing, which was produced from output generated by the Simon Ruffle consultancy, is shown in *Figure 7.6*.

Hidden-surface pictures suffer from the same problems of lack of detail and finish and for the same reasons. Further, because of the poor resolution of the screen, the amount of detail on the output may be degraded even more. Subjectively, however, a half-tone image is much more satisfactory than a line drawing and therefore will often be considered adequate for presentation where a plotted result would not.

Another problem that may be present with computer-generated output is that curved surfaces will be approximated by a number of plane facets. Most programs break down curved surfaces in this way for ease of calculation, but this means that such surfaces exacerbate the storage problem as each plane is held separately. Worse, it means that the quality of the output will be marred, with domes having an ugly tessellated look and columns having a striped appearance. The perspective of the dome of St. Paul's Cathedral in *Figure 7.7* illustrates this problem.

Hidden-surface programs also often suffer from this problem, but the more advanced ones are able to blend the facets together with ingenious shading techniques so as to give a smooth appearance.

Data collection

Early visualisation programs required the user to calculate and type in the spatial coordinates of all corner points on the scene[84]. This was extremely tedious and error-prone and so the programs were seldom used in practice. The modern equivalents have sophisticated data-collection techniques, often requiring much more space in the computer than that needed to calculate the perspectives themselves.

Despite the improvement, the basic problem of computer visualisation techniques remains the effort of collecting the data to define the scene. This takes so long that it will almost always take longer than the manual drawing of a single view. So computer visualisation is not usually cost-effective if only one or two views are required. One survey has found that the break-even point between manual and computer methods lies at about three or four views[85].

This figure obviously varies with the scene. For example, a large housing estate might contain only a few different types of house. Manually, a view of the estate would require each house to be drawn separately, but a set of computer data need only define each type of house once then give the positions at which it is found.

If a building contains a lot of repetition, it may also be possible to define it very economically. *Figure 7.8* shows a view of King's College Chapel, originally produced by Applied Research of Cambridge Ltd. As most of the window bays are identical, only one prototype was needed; the computer was instructed to reuse the

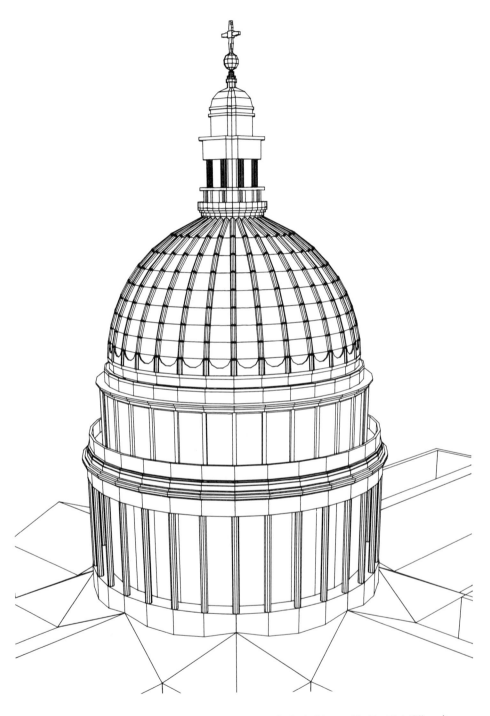

Figure 7.7 A computer-produced perspective of St. Paul's Cathedral (assembled by Nick Gibson)

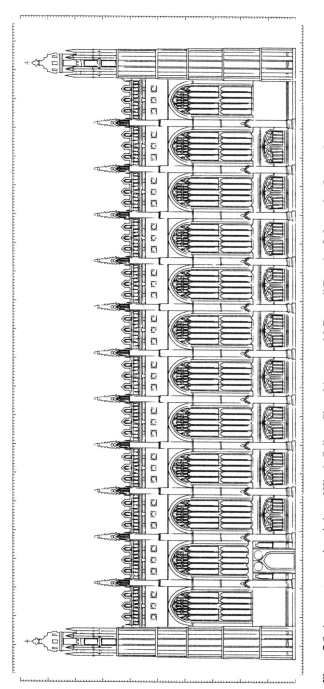

Figure 7.8 A computer-produced view of King's College Chapel (courtesy McDonnell Douglas Information Systems)

same data along the facade. Furthermore, the bays themselves are symmetrical about the vertical axis, so it was finally only necessary to prepare data for half of one bay, and this was automatically mirrored to represent the other half.

Certain unusually-shaped buildings can be represented much more easily by computer methods than manual, such as the geodesic domes popularised by Buckminster Fuller. An accurate manual perspective of such a dome requires long calculation but as the structure is so formalised the data defining it to the computer need only occupy a few lines.

At the other extreme it may well be that if the building contains a lot of arbitrary massing or detailing, the effort of data preparation may be so high as to make it almost impossible for computer visualisation to be cost-effective. As an example we might take St. Pancras Station, designed by George Gilbert Scott, which is pictured in *Figure 7.9*. This building includes many different spires, each finely detailed. The amount of data required to represent it would be considerable.

Figure 7.9 St Pancras station

It is also possible that a building is difficult to describe to the computer not because it is particularly complex, but because of its form. An example of this is the Guggenheim Museum in New York, designed by Frank Lloyd Wright. One view of this is shown in *Figure 7.10*. It would be very difficult to describe the doubly curved shape of this building and would require a lot of data to be assembled.

Figure 7.10 The Guggenheim museum (courtesy Sandra Wilson)

The most popular method of data collection used by modern visualisation programs is to make available a range of elementary solids such as boxes, cylinders and pyramids. The user's data then dimensions these shapes and fits them together, so constructing a block model in much the same way as a child uses a toy building set to build up complex models from simple shapes. An example of this is shown in *Figure 7.11*. A simple representation of a church with a steeple can be exploded into a box shape for the nave; a differently proportioned box for the tower and a pyramid for the spire.

The use of shapes conveniently avoids the tedious and error-prone calculation of the coordinates of each corner on the building. The data is much easier to collect and can be understood at a glance. Shapes can be modified or reassembled if required to represent an evolving design. Curved surfaces such as those on columns and domes are dealt with merely by providing cylindrical and hemispherical solids which have plane facet approximations already applied.

Many perspective programs using this method of data collection allow shapes to be positioned so that they interpenetrate. This can save a lot of trouble with junction conditions. For example, two pitched roofs meeting as shown in *Figure 7.12* can be defined as simple solids with triangular cross-sections interpenetrating as indicated by the dotted lines. If interpenetration were not possible, or if the program did not correctly calculate the lines of intersection, an irregular junction shape would have to be specially defined.

Even greater savings can be made with more complex objects, especially those containing curved surfaces where the lines of intersection will be complicated. Some programs allow the concept of interpenetration to be taken a step further and

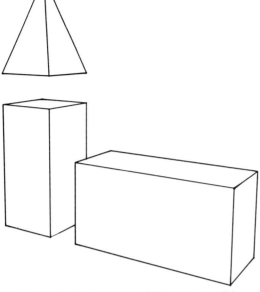

Figure 7.11 Defining a church with elementary solids

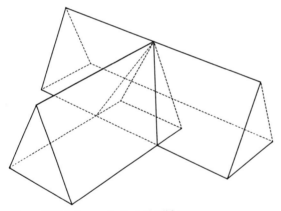

Figure 7.12 Interpenetration of solids

enable the user to specify that one of the solids be 'subtracted' from the assembly rather than added to it. The second solid is regarded as being the absence of material and will cut a hole where it penetrates. An application of this is shown in *Figure 7.13* where two cylinders penetrate, giving the result of a column with a circular hole drilled into it. This result would be almost impossible to achieve without such a facility.

A problem with assembling solids is that if two or more are adjacent, there may be redundant lines where they join. For example the flight of steps in *Figure 7.14* is easily assembled by stacking rectangular slabs, but parts of some lines, as shown dotted, are redundant. Most programs have no facilities for detecting that two surfaces are coplanar and eliminating the boundary lines.

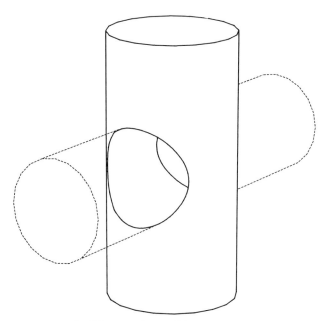

Figure 7.13 The difference of two solids

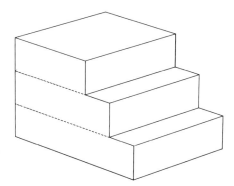

Figure 7.14 Redundant lines in solids

All serious programs have the important ability to reuse a complex shape once it has been constructed. For instance, if a building has several entrances, the flight of steps described above could be defined as a standard segment and located where necessary in the same way as an elementary solid.

Segments can be scaled up and down about any axis as well as moved and rotated, so the flight of steps could be of different width for a wider entrance, or of different pitch or tread, or a combination of these. A scaling factor of minus one forms a mirror image of a segment. This is useful if a building or part of a building shows symmetry; a module of one half can be constructed then joined with its mirror image to form the complete shape.

The assembly of a limited set of standard solids obviously cannot define all possible scenes and for this reason it must be possible to create arbitrary solids for specific tasks. The most useful and straightforward of these is the prism. The user defines a cross-section then 'extrudes' it through the perpendicular axis to form a solid. Such a solid is known as a 'prism'. *Figure 7.15* shows the plan of a building that has been made into a prism.

Another method of creating a solid is to revolve a cross-section about an axis. *Figure 7.16* shows a circular duct elbow piece that was created by sweeping a circle through ninety degrees.

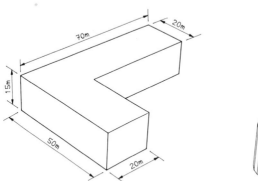

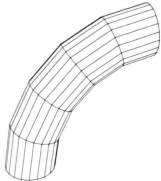

Figure 7.15 A prism   **Figure 7.16** A solid of revolution

It must also be possible to create arbitrary surfaces. For example, a site plan is defined by spot heights, or more rarely in practice by contour lines. It is often necessary to be able to model such a shape. A common solution is that a separate program is provided that takes such data and generates a solid that forms a base for the building.

The actual assembly of solids may be done in different ways. One popular method is to provide a special command language in which an instruction will specify a shape, dimension it and locate it in space. This is easy to use and to understand, but a special-purpose language is usually so specific to its job that it cannot utilise the other powers of the computer and so can make data collection more difficult.

Another method is to provide a set of sub-programs that can be invoked from a high-level language, usually FORTRAN. Invoking any sub-program will generate a specific shape but additionally the powers of the language can be used to generate shapes according to a formula or a set of rules. It is also possible to incorporate decision-making and consistency in checking capabilities to ensure, for example, that all solids above ground level are supported by other shapes and so on. The sub-program solution is more powerful and flexible but does of course require an experienced programmer.

It is very useful in this application to be able to work interactively. Visualising an assembly of solids in three dimensions is not easy. Non-interactive methods require the user to submit a complete set of data and then produce a result, which usually contains mistakes. With interactive working, if a shape is defined or positioned wrongly the error is immediately obvious and can be corrected. A survey found that one interactive program could define a complex office block in one-twentieth of the time required by a non-interactive language[85].

Creating and assembling solids is the usual method of data collection for visualisation programs. There have, however, been experiments based on different principles. One approach under development is the automatic generation of a three-dimensional model from conventional orthographic drawings[86].

Various 'three-dimensional' digitisers can be obtained which can be used to trace over a model by hand. Obviously, some sort of model must exist before this system can be used.

More recently, data has been assembled by using 'intelligent' building blocks. A set of small blocks is provided, each containing a microprocessor that can detect where it is positioned and signal the result to the computer[87]. The user can therefore build up the building physically from the blocks and have the design recorded immediately. This approach is interesting and appears promising, but has had little commercial use as yet.

At a basic level, visualisation programs are classified by the way they hold the data. There are two systems: constructive solid geometry and boundary representation[88, 89]. Boundary representation is the older and the more common system. In this, a scene is defined in terms of the separate plane facets that make up its outer surface. Constructive solid geometry systems store the separate primitive solids that make up the scene, including the 'subtractive' solids that drill holes. Both systems have their advantages and drawbacks. The constructive solid geometry systems use less storage, but are slower in processing as they have to calculate the faceting of curved surfaces and the lines of intersection each time. Constructive solid geometry systems are most popular in engineering applications as it is much easier to calculate things such as mass and volume using this method of data storage[90].

The CAPITOL Program

CAPITOL is a flexible visualisation system written by the firm of Graphicsaid and distributed by the firm of RMJM IT[91]. It is available on a range of microcomputers, including the IBM PC, the Apple IIe, the ACT Sirius and the ACT Apricot. In most cases the computer will need to be fitted with a graphics processor circuit board. The program works interactively and the user can monitor constantly from any angle the building up of the picture on the screen. Any mistakes can therefore be corrected on the spot.

The designers of the system have devised an ingenious solution to the dilemma that generating hidden-line or hidden-surface results on a microcomputer can take many hours, but wire-line results are visually unsatisfactory. The solution is that the output is wire-line views, but the system provides fast methods of colouring areas of the result and of rubbing out selected lines on screen so as to obtain the correct effect. This combination of the computer and human intelligence enables finished views to be obtained at optimal time and cost.

The complete system consists of six programs. Two of them, DIGITIZE and SKETCH, are principally for the definition of shape data; typically building forms or ground shapes. DIGITIZE is used with a digitising tablet and allows the user to trace over existing drawings in order to input data. SKETCH performs a similar function, but the user draws lines on the screen with a cursor or by typing line lengths. A drawing created by either of these programs can be saved away as a file on disc.

DIGITIZE has the very advanced capability of being able to define three-dimensional data completely interactively. A plan and elevation are fixed to the digitising tablet, then each line end is defined by two point indications: one on each view. The results are shown on the screen as an isometric view, so the user can see the scene build up. This elimination of typing in dimension values and the ability to see the effects of commands as they are issued makes data definition fast and easy.

The main program in the system is called CREATE. This program is used to create and assemble solids and generate the perspectives or other views. The maximum complexity of the scene that can be supported varies with the amount of computer memory available: on a 896 kilobyte machine the limit will be around 5,000 planes, which is reasonably high.

CREATE uses a set of 41 English language commands which are abbreviated to their first two characters. The first step on entering the program is usually to load a data file. This is done by the 'RESTORE' command followed by the name of a file, which may have been created by DIGITIZE or SKETCH. If the user has previously digitised a plan shape, for instance, and filed it under the name PLAN, the command:-

RE   PLAN

will load the data into memory.

One of the most useful commands is 'EXTRUDE'. This command takes a plan or cross-section and extrudes it through the third dimension to form a solid. The command is followed by three numbers, giving the amount of extrusion along the three axes. With a plan, the extrusion should be done along the third axis (top to bottom) only. As an example, the plan shape described above could be loaded and turned into a block model of a building 30 metres high by the commands:-

RE   PLAN
EX   0   0   30

Plans or cross-sections must have curved lines approximated by straight line segments. This is automatically necessary as DIGITIZE and SKETCH only deal in straight lines. If therefore a plan contains a circular column, it will in fact be stored as a polygon. When extruded, this will give the sort of effect illustrated in *Figure 7.17*.

The EXTRUDE command is the only one that will generate a solid from two-dimensional information. However, it is possible to transform solids to make other shapes and to store them away as disc files for future use. Most firms will quickly build up a library of useful shapes. Various standard shapes are supplied with the system. The most commonly used of these is called 'CUBE' and contains a standard cube shape of 100mm on each side. To load the cube, the user types:-

RE   CUBE

The most basic transformation command is 'SCALE'. This is followed by three numbers which scales the solid along each axis. Thus to turn the standard cube into a box representing a 300 mm square structural column of 3 metres height, the commands will be:-

RE   CUBE
SC   3   3   30

The 'WEDGE' command allows the user to 'taper' solids. The command is followed by two numbers, both representing points on the x-axis. Lines parallel to

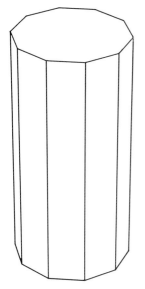

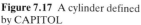

Figure 7.17 A cylinder defined
by CAPITOL

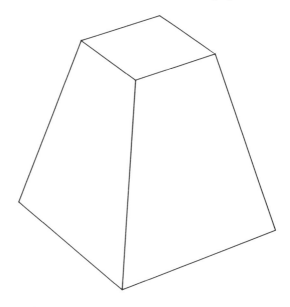

Figure 7.18 An Aztek pyramid defined by CAPITOL

the x-axis in plan taper towards the first point and lines parallel to the x-axis in the vertical plane towards the second.

This command is surprisingly flexible in use and can produce a variety of pyramidal shapes by varying the taper. For example, a flat-topped 'Aztec' pyramid as shown in *Figure 7.18* could be produced by:-

```
RE        CUBE
WE        200       200
```

An 'Egyptian' pyramid can be produced by having lines taper towards a point at the top of the cube. Thus the pyramid in *Figure 7.19* is produced by:-

```
RE        CUBE
WE        100       100
```

If the two points in the WEDGE command are different, one taper will be sharper than another. So a typical hipped-end roof as shown in *Figure 7.20* could be defined by loading a cube, scaling it to the correct proportions, then tapering it, as follows:-

```
RE        CUBE
SC        1      1      2
WE        100      300
```

If either of the two points is at infinity on the x-axis, then no tapering will occur. A value of 10,000,000 is effectively infinity, so a gable-end roof as shown in *Figure 7.21* could be defined by:-

```
RE        CUBE
SC        1      1      2
WE        100      10000000
```

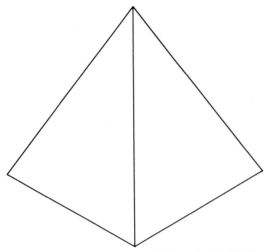

Figure 7.19 An Egyptian pyramid defined by CAPITOL

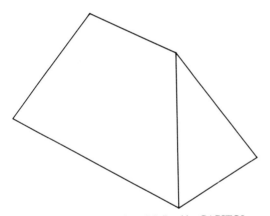

Figure 7.20 A hipped-end roof defined by CAPITOL

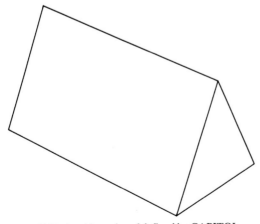

Figure 7.21 A gable-end roof defined by CAPITOL

By moving the cube off-centre before tapering, different results can be obtained. One useful example is a simple wedge shape, which can be used, for instance, as a car ramp. Other effects can be produced from different solids, so a cylinder could be converted to a cone, say.

Having defined the elementary solids, they can be assembled into complex scenes. The most basic assembly commands are those to move solids around. These are 'MOVE' and 'MOVE TO' (abbreviated to 'MT'). Both commands are followed by three numbers representing a point in space. The difference is that MOVE moves the solid relative to its original location while MOVE TO moves to a fixed point whatever the original location was. Thus stacking one cube on top of another, as shown in *Figure 7.22*, can be done by:-

```
RE        CUBE
RE        CUBE
MO        0      0       100
```

It is also possible to move solids interactively. The command 'FLOAT' allows the use of the cursor to position the solid by eye on the screen.

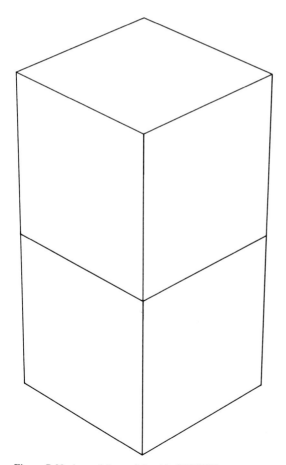

Figure 7.22 Assembling solids with CAPITOL

It should be noted here that all operations act upon the last solid that was loaded by a RESTORE command. Thus the MOVE command in the last example only moved the second cube. However it is possible, and often extremely useful, to be able to operate upon a group of solids or upon the entire scene. The 'OBJECT' command allows the user to select such groups and then commands will index all the data selected.

Another important assembly command is 'ROTATE' which is followed by three numbers specifying rotations about the three possible axes. Thus a cylinder could be extruded from a circle, then turned on its side to form a horizontal pipe, for example.

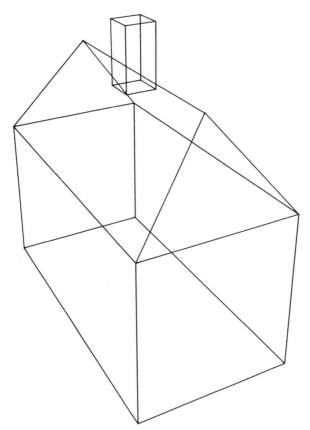

Figure 7.23 Assembling a house with CAPITOL

An example of a simple assembly might be a basic house shape. A house could be defined as a box surmounted by a pyramid for the pitched roof. The chimney could be represented by a smaller box that pierces the roof. This house is illustrated in *Figure 7.23*. Note that a disadvantage of CAPITOL is that the lines of intersection between the roof and the chimney are not calculated.

The steps are to call up a cube and dimension it to form the house body. Then call up another cube, dimension it and taper it to form the roof and move it into position on top of the previous shape. Finally, a third cube is called, dimensioned and moved to form the chimney. The exact commands could be:-

RE	CUBE		
SC	50	100	50
RE	CUBE		
SC	50	100	30
WE	3000	12000	
MO	0	0	5000
RE	CUBE		
SC	10	10	20
MO	0	0	7500

Assembled shapes are most useful when they can be used repetitively. Our house shape might be used dozens of times in different positions to form an extensive housing estate. To preserve a shape for re-use it must be saved away as a disc file. The user must first use the 'OBJECT' command to select all the shapes. The form:-

OB   ALL

selects all elements in the scene. Then the 'SAVE' command followed by the name of a disc file will create a disc file containing the current data. In the current example:-

SA   HOUSE

sets up a disc file containing the house shape. This file is now an elementary segment in itself and can be called by the RESTORE command and positioned as many times as required to build up the housing estate.

The complete scene can be built up easily by using such commands as described as well as a range of others which are provided for various situations. The user can take different views of the scene at any time to monitor input or to generate outputs. Various commands are provided to move the viewpoint and the line of view. The 'VIEW' command is used to generate a three-point perspective on screen. The program also uses certain function keys on the keyboard to provide options of axonometric, isometric and orthogonal projections and to display the scene in plan or elevation.

All results will be wire-line views. These will often be acceptable in themselves, but the user is also provided with facilities to change them into hidden-line or hidden-surface effects. Hidden-line views are obtained through the use of the ERASE program which can remove lines or portions of lines after the view has been obtained through the CREATE program. The results can then be drawn on a pen plotter. As each line has to be handled individually this program is inevitably rather slow in use.

More advanced facilities are provided to generate hidden-surface effects. The 'PLOT' command within the CREATE program is provided to accomplish this. In essence, the user selects a colour, then indicates a single point on the image. That colour will flood out from the point, stopping when it reaches a line drawn on screen. In use, it is very fast to move the cursor about the image indicating points on the view, so painting in the surfaces quickly and exactly. *Figure 7.24* shows how this might be done for a simple cube.

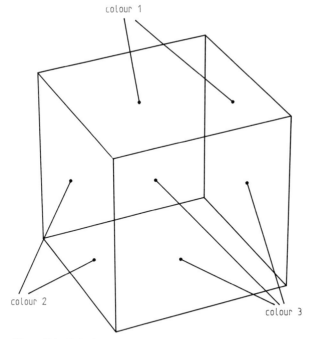

Figure 7.24 Painting a cube perspective with CAPITOL

This method of working is made more powerful by the option of specifying that painting must stop only at a line of a given colour. If the scene is assembled of solids in differing colours, therefore, many surfaces can be painted in with a single point indication even though they overlap with surfaces on other solids.

Chapter 8

Job management

The need for job-management systems

During the construction process the contractor must control large numbers of men, large quantities of materials and a great deal of plant and machinery. All these resources are expensive, and if the contractor is to remain competitive they must be tightly and efficiently controlled[92]. By contrast, during the design process only a handful of architects and technicians are employed, and they use virtually nothing in the way of plant and materials beyond drawing instruments and paper. As so few resources are employed, control is relatively simple and does not have to be particularly strict. However, buildings are getting larger and more complex and require bigger design teams. There is strong pressure from clients to finish the building as early as possible. Both these factors make project control more difficult. In addition, the costs of employing staff are still rising and the office that does not make the fullest use of its employees will find its already narrow profit margins eroded still further.

A typical control system will specify when tasks should be carried out and how many men will be needed at each stage of the project. It should also be able to reschedule these tasks to take account of the number of staff actually available and their specialist skills. Such techniques will allow the administrator to look ahead and prepare for each task as it falls due, and to specify deadlines accurately. This should help to cut down on the last-minute rushes and late-night working that happen so frequently in many offices and must harm the quality of the design. The administrator will also be able to keep staff fully employed by organising transfers between jobs at different stages and by knowing when to employ extra staff and when to refrain from replacing leaving staff. This last advantage is especially important in an architect's office as professional staff do not expect to be hired and fired with the fluctuating workload, and temporary staff are expensive and require training before they can be of greatest use.

Another important benefit of formal control systems is that they will pinpoint the 'critical' tasks that must be finished on time if the completion date is not to be delayed. Thus the administrator can make sure that these activities start on time and have sufficient staff, if necessary by delaying the completion of non-critical activities. Conversely, it is only by speeding up the critical tasks that the completion date can be brought forward. Therefore if the project is running late or needs to be hastened for some reason, attention can be accurately focused on the most relevant activities and not wasted on tasks that have leeway in their completion time.

The sort of results described can also be obtained by a seasoned administrator on the basis of knowledge and experience, and if such a person is available it may well be more efficient to control the project manually rather than use a computer program. However, if a sufficiently experienced person is not available, use of the computer can ensure adequate results and avoid gross errors and oversights. This is a good example of the way a computer can 'de-skill' fields previously needing years to master. An expert in almost any technical subject can program a computer to follow the best procedure in every combination of circumstances and to evaluate the most accurate formulae, however laborious. The expert's experience is then available to anyone with access to the program, who generally needs only to be familiar with the broad outlines of the subject to achieve good results.

Principles of critical-path techniques

The control technique most commonly used is the 'critical path' method, although there are others. This method of project analysis has been around since 1956 and is in widespread use to control all sorts of complex activities from building to the space programme. It can be applied by manual methods, and in fact this was often done before computers became commonly available, but because of the large number of laborious, if elementary, calculations involved it is much more suitable for computer evaluation.

The critical-path method relies on two basic assumptions: first, that the task to be analysed can be split into a number of distinct smaller tasks and secondly that, apart from the first activity, each task cannot start until certain of the others have been completed. These assumptions are not always valid, as will be seen, but most projects can be expressed in this way with relatively small amounts of distortion and can then be analysed in a number of useful ways.

As a very simple example, we can take part of the construction process: the task of painting a room. This can be broken down in many ways at all levels of detail. For example, a contractor would probably find it most useful to specify painting a room as a single task to be carried out by one man, whereas a time-and-motion investigator might require a detailed analysis down to the level at which taking the lid off a paint pot is a separate activity. For our purposes, we can divide the task into: (1) painting the ceiling, taking four hours; (2) painting the walls, taking eight hours; (3) preparing the woodwork, such as the skirting boards and window and door frames, taking two hours; (4) painting the woodwork, taking three hours; and (5) a final inspection that takes one hour.

Having determined the best division of the task into sub-tasks, the correct order for the activities must now be decided. The first constraint is clearly that preparing the woodwork must precede painting it. The next is that the ceiling should be finished before the walls or woodwork are started, to avoid the risk of splashing completed work. Lastly, the final inspection must follow the completion of the painting of the wall, ceiling and woodwork. This information can be expressed as a diagram, using boxes for each activity and linking arrows to represent the order in which they are carried out as in *Figure 8.1*.

Note that no link has been drawn between painting the ceiling and final inspection, although it was one of the constraints. Such a link is unnecessary because painting the ceiling precedes painting the other surfaces and they precede final inspection. It would not be wrong to draw in the link but it saves computer time not to, and makes the diagram clearer.

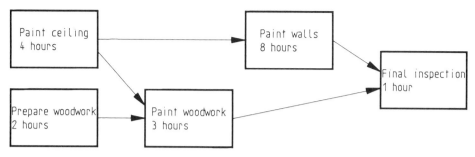

Figure 8.1 A precedence diagram

This network can now be analysed, and it can be calculated that the total length of time for the project will be thirteen hours. Also, the activity of painting the woodwork can start at any time between the fifth hour and the tenth hour of the project without delaying its completion. The usual terminology is that this activity has five hours 'float'. However, the activities of painting the ceiling, painting the walls and the final inspection must start as soon as the preceding activities are finished and must not take longer than their scheduled times if the project is not to over-run. These activities are said to be 'critical' and to lie on the 'critical path', i.e. the longest path through the network.

The activity of preparing the woodwork is a little different. It could be delayed by up to seven hours from the start of the project without delaying the completion time, that is it has seven hours float. But if full advantage is taken of this float, then painting the woodwork becomes critical; it must start as soon as the woodwork has been prepared and must not be delayed. If the activity of painting the woodwork is not to be interfered with and is to retain all its float, then preparing the woodwork can only be delayed by up to two hours. Thus although its 'total float' is seven hours, its 'independent float' is only two hours.

The diagram we have been using is drawn with the activities given prominence and is called a 'precedence diagram' or 'activity-orientated diagram'[93]. It is also possible to draw the network giving prominence to each distinct point in time. With this method, the points at which activities start and finish are usually represented by circles in which start and finish times can be written in later. The activities are then shown as arrowed lines which indicate the relationships between the points in time. To illustrate this we can draw out the room-painting project in this form, known as an 'arrow diagram' or 'event-orientated diagram'. The arrow diagram is shown in *Figure 8.2*.

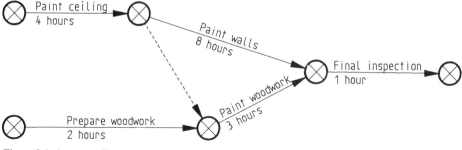

Figure 8.2 An arrow diagram

Note that in this form a dotted line has had to be included to indicate a relationship where an activity links to more than one following activity. This line is regarded as a 'dummy activity' having zero duration.

Arrow diagrams are neither as clear nor as easy to construct as precedence diagrams, and are in fact being supplanted. However, they were used for many years before the introduction of precedence diagrams and are still in widespread use[94].

So far we have only considered relationships where one activity commences on the completion of others: a finish-to-start relationship. There are others, however. For example, when mixing mortar, the ingredients should not be put into the drum of the mixing machine before it starts operating or premature setting may take place. This is a 'start-to-start' relationship, where the start of one activity cannot precede the start of another. Using the same example, mixing of the ingredients obviously cannot continue after the machine stops operating. This is a 'finish-to-finish' relationship where the finish of one activity cannot follow the finish of another. The process is shown as a precedence diagram in *Figure 8.3*.

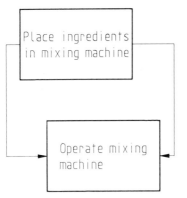

Figure 8.3 Start-to-start and finish-to-finish links

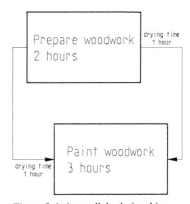

Figure 8.4 A parallel relationship

More frequently in practice, one activity is dependent on the start of another after some delay. For instance, all the woodwork does not have to be prepared before it can be painted; provided enough men are present, one man can fill and prime the woodwork and another can paint it after it has dried, say one hour later. But in this case there is also the constraint that painting the woodwork cannot finish until at least one hour after the preparation of the woodwork has finished. This situation is frequently encountered; one activity runs parallel to and lagging behind another. This is illustrated in *Figure 8.4*.

Some critical path programs allow delay times for, say, postage time, or setting and curing times to be placed on the linkage arrows as shown, but others require that the delays be shown as separate 'activities'.

Critical path techniques impose a formalism in specifying activities and their relationships that is not always present in practice. It is, for example, impossible to specify properly a process of feedback or continuous refinement. An architect often works by scribbling a design, then redrafting it in a slightly changed or more detailed form. This process is continued until a satisfactory result is obtained. The only ways that this can be specified by the critical-path method are either as a single activity, or as a fixed number of steps. There is no way to make one activity loop back into itself or into any other activity that has already commenced.

Another rather unnatural restriction is that it is only possible to express links between activities relative to their starting and finishing points. It is not possible, for instance, to show directly the architect sending a consultant information halfway through the production of the working drawings. Instead, either the activity of producing working drawings must be split into two activities with a link to the consultant's activities at the end of the first stage, or there must be a start-to-start link between the architect's and the consultant's activities with a delay time on the link to give the effect of a later start to the following activities.

Construction of a network is greatly simplified by drawing out the diagram of activities and their relationships first. The whole project can then be seen in a pictorial form and loops or omissions and other mistakes are much easier to detect. Some programs allow the network to be drawn on the plotter from the input data, in which case it is not necessary to draw the initial network very carefully. Otherwise, it will be necessary to draw neatly and legibly and to lay out the activities with plenty of space between them to allow for the inevitable future alterations. When changes are required, because of a change in policy or because of a hold-up during the project that affects subsequent activities, it is usually essential to be able to refer back to the network diagram to understand the implications of the alteration.

Construction of the network can also be simplified by the use of composite activities. This entails replacing a single activity in the network by a number of activities forming a sub-network of their own. For instance, a network might contain an activity entitled 'Prepare production documentation' which can obviously be replaced by a network of its own. This facility is provided by many critical-path programs. It means that a skeleton network can be defined at a general level by the project manager, then the activities can be replaced by detailed networks, prepared in isolation by the appropriate specialists, to give the final complete set of data.

Practice is needed to construct good critical-path networks. The user has to train himself to express reality in the formalised critical-path conventions, and has to decide on the level of detail of the network; too much detail will be confusing while too little will be unhelpful. The way the activities are linked can also be a problem; putting in links that are not strictly necessary will make the project appear too rigid and lacking in float time, while putting in too few links may give the impression of more leeway than there actually is. The technique is not difficult to master, however, and the amount of information to be collected is not high. It will usually be possible to prepare the data for a typical job in a day or two.

The data preparation time can be reduced to a few minutes if an existing network can be used. This could be a network from a previous similar job that can be used again with small modifications, or it could be a standard network that is applicable to a range of jobs. Various guides to the management of building construction have been published, for example, the RIBA Architect's Job Book[95], and it is possible to construct networks based around such guides. Most offices have a preferred way of organising jobs and these policies can be embodied in a standard network and are then automatically used when applied to a scheme.

The data preparation does not end with the specification and ordering of the tasks to be performed. This is because without further constraints it is likely that the logic of the network will allow many activities to go on at the same time, even if there are not enough staff available to carry them all at once. For this reason, another factor must be introduced into the critical-path analysis: the allocation of resources. Resources are most often people possessing the skills required to carry out the tasks in question, but they can also be materials, or even money. For example, a contractor will only have a certain number of bricklayers available and this will set a limit on the number of bricks that can be laid in the times specified. The contractor may only be able to take delivery of a certain number of bricks in any given week and this also sets a limit to the amount of bricklaying that can be done. In this case both bricklayers and bricks represent resources associated with certain activities.

Critical-path programs allow the user to specify the resources required for each activity and also the total number of resources available. When scheduling the activities, the program will shift the starting dates of activities within their float times, or if necessary increase the time of the project to accommodate the activities within the resource limits. As well as ensuring that these limits are not exceeded, the program will also attempt to level out the use of resources so that, for example, the same number of draughtsmen are used throughout the entire project. This makes consistent use of resources and simplifies management. The program can also report on what resources are required and when, the number of staff needed, the right time to order the minimum quantities of materials and so on.

Some of the more versatile programs allow a variable amount of resources to be utilised over the period of a single activity. For example, painting might require one man for the first day to make preparations, one man on the last day to touch up mistakes and five men for the intervening period. This obviously gives more accurate results and avoids splitting that activity into three parts as less sophisticated programs would have to. With most programs, the limits on the resources can be varied over the project period. Some programs allow two limits to be specified, the first a 'normal' maximum and the second a limit that could be met by overtime working, taking on temporary staff, paying extra for earlier delivery dates of materials, and other expenditure necessary to meet deadlines.

When resource allocation is provided, it is very easy to associate costs with each resource. Thus bricklayers are paid a certain amount each week; painters a different amount; cement costs so much per cubic metre and so on. Comparatively little extra information need be supplied as there are not usually many different resources. Given these unit costs and the resources needed for each activity, the program can easily calculate the total cost of the project, so that it can be checked that the budget will not be exceeded. The program can also print out the costs that will be incurred month by month; this is important when money is tight and cash flow is a problem, or at a time of high interest rates. As well as the above information, the user must also provide a short set of data specifying the start date of the whole project, holiday periods and other basic information.

Many job-management systems allow the user to force scheduled starting dates for some activities. Deadlines can thus be incorporated into the network and the float times for the preceding activities will automatically adjust themselves. One useful application of the facility is to specify a finishing date for the project so that, when the network is processed, the appropriate starting date will be calculated. The problem with scheduled dates is that it will sometimes be impossible to meet them given the network specification. This is indicated by the float times becoming negative in one or more cases. A float time of minus one day, for instance, means that at least a day has to be saved somewhere to meet the deadlines.

A fault of many job management programs, so far as the architect is concerned, is that they are too complex. A major construction project may involve thousands of separate activities that the main contractor and sub-contractor must carry out. Projects of this size require management specialists and computer programs that are large and flexible. In a typical architectural practice, the situation is quite different. A building design programme can be adequately defined by a hundred or so separate activities and rarely requires as many as a dozen architects and technicians. The material resources used are negligible. Very few offices will be able to justify a project manager and most of the advanced facilities provided by the largest job management programs will not be required. It is therefore better for the architect to choose a relatively simple and easy-to-use program rather than a very comprehensive one that will create more problems than it solves[96].

Outputs from job-management programs

A wide variety of outputs is possible from most job-management programs. Perhaps the most familiar, and certainly one of the most useful, is the barchart or Gantt chart. This represents in pictorial form the time-span of the tasks to be carried out and can show deadlines very clearly. The conventional practice of colouring the activities or proportion of activities that have been completed gives an immediate impression of the progress of a project.

The barchart that was illustrated in *Figure 2.10* is a typical example of this form of output as shown on a computer screen. The dark-coloured bars are the critical activities and the light the non-critical. These latter have their float times shown by thinner lines extending from the bars.

The activities on the barchart, and on most other forms of output, can be ordered in a number of ways. The most common is the order of their starting date. This is convenient because the next tasks to be undertaken are obvious. Other orderings can be useful in different circumstances. For example, ordering by finishing dates shows immediately which tasks should be completed by a certain time, ordering by length of float time puts at the top of the list the activities that must be most precisely controlled; and leaving the order of output the same as the input data allows quick correlation with the network diagram. It is also usually possible to specify that certain activities belong to different groups and to split up the barchart or other output between these groups. This is most commonly done by trades, so that the architect's activities will form one separate section, a given consultant's activities another and so on.

All outputs can be produced over a limited time-span if required. A typical output might omit activities completed before the current date and only stretch for six months from the current date. This focuses attention on the immediate situation and is often less confusing than presenting the entire project.

A more convenient form of output for desk work is a list of activities with their important calendar dates. The example in *Figure 8.5* shows the earliest starting and latest finishing dates for each activity and, if the activities are not critical, the latest starting and finishing dates each one can have without interfering with subsequent float times.

This output is more compact than a barchart and precise dates can be noted more easily than by reference to the date scale on the barchart; it does not, of course, give such an immediate impression.

A SIMPLE DESIGN PROCESS

CODE	ACTIVITY DESCRIPTION	TIME	EARLY START	LATE FINISH	INDEPENDENT START	INDEPENDENT FINISH
ARCH01	PREPARE SKETCH PLANS	10	THU 2 JAN 86	WED 15 JAN 86		
CLNT01	CLIENT APPROVAL	3	THU 16 JAN 86	MON 20 JAN 86		
ARCH02	PREPARE 1:200 FLOOR PLANS	20	TUE 21 JAN 86	TUE 25 FEB 86	THU 23 JAN 86	WED 19 FEB 86
ARCH03	PREPARE 1:50 ROOM LAYOUTS	40	TUE 21 JAN 86	MON 17 MAR 86		
CONS01	CONSULTANTS DEVELOP SCHEME	20	THU 20 FEB 86	THU 27 MAR 86	FRI 28 FEB 86	THU 27 MAR 86
COST01	COST CONSULTANT COSTS SCHEME	10	TUE 18 MAR 86	MON 31 MAR 86		
ARCH04	FINAL PRESENTATION	1	TUE 1 APR 86	TUE 1 APR 86		

Figure 8.5 A calendar report

One extremely useful form of output that unfortunately is not found in many critical-path programs is the ability to redraw the network diagram on the computer plotter. Network diagrams take a long time to draw initially because it is not clear how the activities should be laid out on the paper. Therefore drawing a neat diagram has to be done in two stages: an initial sketch to see the organisation and a complete redraughting with the activities in order of starting times and with the minimum of links crossing to give the clearest impression. Several initial drafts of a complex network may be necessary and others may be required during the life of the project.

Thus a network-drawing facility can save a lot of time. A portion of an output from such a program is shown in *Figure 8.6* and illustrates the detail design stage of building construction.

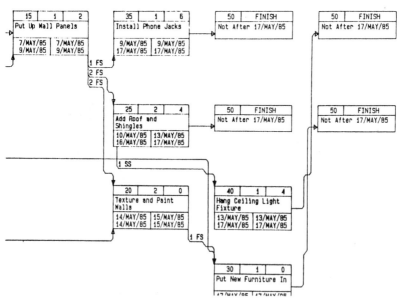

Figure 8.6 A network output (courtesy Abtex Software Ltd)

Defining and correcting the network can be time-consuming even on a computer, and the best programs provide sophisticated facilities for this task. One such program is 'PERTMASTER', which is written and distributed by Abtex Software Ltd.[97] *Figure 8.7* shows a user 'zoomed in' to a single activity in a network in order to make alterations.

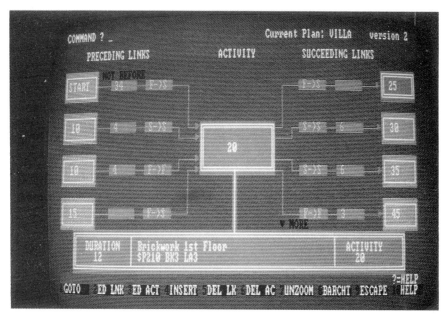

Figure 8.7 A zoomed-in portion of a network (courtesy Abtex Software Ltd)

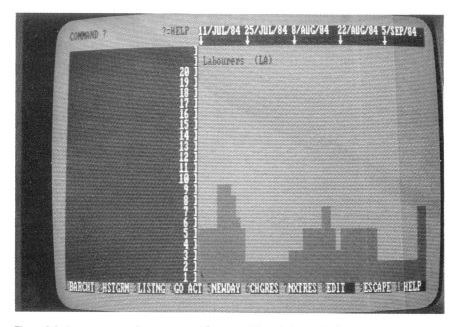

Figure 8.8 A resource requirement output (courtesy Abtex Software Ltd)

Another important form of output is one showing the use of resources and the costs included in the project. One of the clearest forms to show these is the histogram. This is a familiar type of data presentation that uses vertical bars of different heights to represent values. By studying the output, a manager can see how many men of what skills will be required in the next time period and so can redeploy the remainder. The manager can also see the costs involved and again can make appropriate arrangements. A typical histogram of resources is shown in *Figure 8.8*. It illustrates the variation in the number of labourers required over the duration of a project.

Data Preparation - a worked example

There are many different network-analysis programs in existence. A survey a few years ago investigated in some detail almost 100 available in the UK, all with minor differences[98]. Because of this proliferation, data preparation will be illustrated here in a generic manner that can, with slight modifications, be adapted to any critical-path analysis program.

Figure 8.9 shows a network of part of the design process, on a much cruder level than would be used in practice.

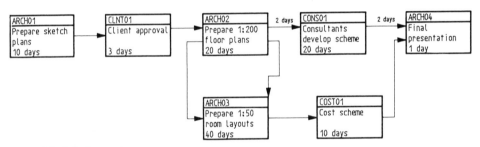

Figure 8.9 A design process

The diagram shows that the first activity is to be the preparation of sketch plans, taking ten days. When completed they are sent to the client for approval, a process expected to take three days. When approval is given, work can start on the floor plans, to take 20 days. The preparation of the large-scale room layout drawings takes place in parallel with the floor plans, but cannot start until the floor plans have started and cannot finish before the floor plans have finished. This activity takes 40 days. The consultants can begin their work when the floor plans are finished, plus two days delay for packing and postage. The consultants' work takes 20 days and there is then another two-day delay before their drawings are back with the architects. The cost consultant or quantity surveyor can cost the schemes after the large-scale room layouts are finished and takes 10 days. When this is completed and when the consultants' drawings have arrived, the final presentation can take place and is to last a single day.

Each activity is given a brief description such as, 'Prepare 1:200 floor plans', so that the outputs will be readily intelligible. It is also given a short identification code of up to six characters, which is used in preparing the data and eliminates typing out the description in full every time an activity is referenced.

The most basic part of the data is the activity-relationship table. This gives the identification code, description and duration of each activity and the code of the activities that follow it in the network. It also specifies the types of link and any delay times. Using the abbreviations 'F', 'S' and '-' to represent finish-to-finish, start-to-start and finish-to-start links respectively, we can write out the network as an activity-relationship table (below) for typing into the computer.

Code	Description	Duration	Following Activities	Link Type	Link Delay
ARCH01	Prepare sketch plans	10	CLNT01	–	0
CLNT01	Client approval	3	ARCH02	–	0
ARCH02	Prepare 1:200 floor plans	20	ARCH03	S	0
			ARCH03	F	0
			CONS01	–	2
ARCH03	Prepare 1:50 room layouts	40	COST01	–	0
CONS01	Consultants develop scheme	10	ARCH04	–	2
COST01	Cost consultant costs scheme	10	ARCH04	–	0
ARCH04	Final presentation	1			

To include resource scheduling, two other sets of data must be prepared: a resource-schedule table to specify the different resources and their limits, and a resource-assignment table that specifies how much of each resource every activity will require. The resource-schedule table, in a similar fashion to the activity-relationship table, gives a description and a code to every resource. It also gives the maximum amount of that resource available, and the time from which that maximum applies. Thus if the available amount of a resource will vary over the project time, this can be specified by putting in several start dates with the appropriate maxima. The resource-schedule table may also have an entry giving the cost of that resource so that cost can be calculated over the project.

It is not necessary to specify all resources, or indeed any at all if resource allocation is not considered helpful enough to be worth the extra effort. In our example, the network is for the job leader who wishes to optimise the use of architectural staff, but is not concerned with the number of staff the client and the consultants must use. The only matter that affects the architect is the time these others need to carry out their parts of the design. Therefore only the activities carried out by the architect need have resources detailed or assigned; the program will then schedule the other activities as if they had no resource restrictions.

For the purposes of our example, we will assume that the office is able to assign to the job one architect, two technicians and a partner of the firm, who will lead the final presentation. All of these staff are available from the start of the project, on day 1. Staff costs per day are expressed in units whereby one day of a partner's time is worth 100 units; an architect's 50 units and a technician's 40 units. The resource schedule table can now be set out as below.

Code	Description	Number Available	Availability Date	Cost/day
ARCHIT	Architect	1	1	50
TECHNI	Technician	2	1	40
PARTNR	Partner	1	1	100

The next block of information to be given is the resource-assignment table that specifies which resources, and in what amounts they are needed to carry out each each activity. For convenience of data preparation, the activity codes and the resource codes are used rather than the full descriptions. The table illustrated has a column simply giving the total quantity of each resource required by any activity. As remarked earlier some programs are actually able to vary the resource quantity over the duration of individual activities.

It might be expected that sketch plan preparation will involve only the architect. The preparation of the 1:200 floor plans is estimated to need an architect assisted by one technician. The preparation of the large-scale room layouts requires two technicians to develop the design and the final presentation needs the architect and the partner. The resource-assignment table can now be set out.

Activity Code	Resource Code	Resource Quantity
ARCH01	ARCHIT	1
ARCH02	ARCHIT	1
	TECHNI	1
ARCH03	TECHNI	2
ARCH04	ARCHIT	1
	PARTNR	1

Obviously there is a problem here. By assigning more manpower the activities can be completed earlier, up to a point. The usual course is for a reasonable balance to be struck between too short a duration with too many men, which is inefficient and can give poor results, and too long a time with too few men, which gives good results but may delay the project. After studying the output from the first computer run the manager can often see that better balances may be struck in certain activities. Those on the critical path can be given more resources and those with a great deal of float can be given less. This insight is one of the advantages of job-management programs and enables the use of resources to be optimised.

As well as this basic data, the user must supply a set of miscellaneous information that applies on a general level. This will include the time units in which the data is prepared, the cost units, the public holidays on which no work will be performed on the project, the project title, its start date and various other matters. The complete set of data can now be presented to the computer program and will give the sort of results that have been illustrated.

The example given has been a very simple one; with so few activities it would not in practice be worth using computer analysis. A realistic network for the design of a building would normally contain about a hundred activities, and at this sort of complexity a formal approach can often give better results than an intuitive one.

Chapter 9

Design aids

Types of design-aid programs

A glance at the articles and conference papers on computers in architecture throughout the 1960s will show a very heavy emphasis on programs that aid the sketch design process. Relatively little attention was paid to such matters as database management, draughting or visualisation. Today, the situation is exactly reversed: much attention is paid to programs that help the later stages in the building process and very little to design-aid programs.

The reason for this is mainly the fall in the cost of computers since that time. Processing time was so expensive that it was clearly impossible to make savings over manual effort in straightforward tasks, but as it is very difficult to put a price on a better design, it was quite possible to justify high computer costs in this application.

Another important reason is that it turned out in practice that computers did not aid the sketch-design process very much. They could often give a certain amount of help, but this was so minor in relation to the total process as hardly to justify the use of an expensive machine.

It is probably fair to say that design-aid programs still offer relatively little to the architect. In certain situations, however, especially when dealing with large buildings where human intuition is less effective, valuable gains can be made by the careful use of these programs.

Broadly speaking, there are two ways of using the computer in design: in a generative way, where a design is produced from the basic input data, and in an analytical way where the user proposes a design and the computer checks its viability. Although the generative approach is one that has received a great deal of attention, it has been almost completely abandoned in its purer forms by practising architects, because the results produced are inadequate. This is because it is not possible to build into a computer program many of the most important factors that must be considered in a design. The analytical approach is now more favoured; if the architect produces a design and the computer checks such aspects of it as it is able to, it is possible to create a powerful symbiosis of man and machine.

The difference between the generative and the analytical approaches can be illustrated at a simple level by the example of window design. If the architect designs a window, the computer can very easily check how much daylight and sunlight it lets in and how much heat it lets out. The design will already have taken into account such factors as the relative proportions of window to wall dimensions needed to give a pleasing impression both inside and outside, and the view from different points in the room. By modifying the design and rechecking, the user can get a good solution in two or three tries.

The equivalent solution with a generative program would be to produce a window design that meets the environmental requirements. There are of course an almost infinite number of possibilities, so the computer is forced to choose a solution which may or may not meet the more subjective criteria.

The difference in the two approaches becomes even more marked when more complicated design problems are considered. Generative programs for producing building plans typically try to minimise circulation. They ignore social customs, the state of mind and body of the user, aesthetics, the activities taking place in the rooms, the view from the windows and so on. This simply means that their results are inadequate: just because a requirement cannot be fed into a computer does not mean that it does not exist. In contrast, an analytical program can test a solution for circulation costs or other objective criteria after the architect has produced a design that takes into account the more intuitive factors. Thus the interaction between man and machine can take advantage of the abilities of each.

The early generative programs did not allow for interaction; they simply went on and produced a design. The more modern versions allow the user to interact and to fix parts of the solution after which the computer finishes the design around those fixed points. This does improve their usefulness.

Analytical design methods can be further divided into dynamic and non-dynamic analysis. Non-dynamic analysis is the simplest and consists essentially of evaluating formulae and checking the rules relating to fixed situations. Evaluating the construction costs of a building is one application: depending on the amount of data given, the computer can add up quantities of materials and give a price. Cut-and-fill analysis is another popular use of non-dynamic analysis; given the terrain map and an outline plan of the building, the computer can calculate its optimal siting so as to minimise the amount of earth moved.

Non-dynamic analysis can be applied to the costing of a design at a number of levels. One aspect of design is the disposition of staff and equipment about a building. This is not a problem to be solved only once: as conditions change, some departments will increase in size and other shrink, and new types of machinery will have to be accomodated. In large buildings on urban sites, a great deal of money is involved as the rental rates are very high and wasted space cannot be tolerated. This problem has been formalised under the name of 'facilities management'[99]. Its main aspect has been that of financial optimisation and for this reason it is usually under the direction of managers with a training in accountancy. However, computer programs now exist which can take a sketch plan and associate it with a database of personnel overheads and equipment costs and from this evaluate the cost of a layout. This means that architects can use their presumably superior design abilities to propose solutions and with the help of the computer ensure that they are not only workable but also economic.

Dynamic analysis, or simulation, is a much more powerful technique than non-dynamic analysis. With this method, a model of the activity in progress can be run on the computer and by studying the consequences of the activity it can be seen if the design is suitable to contain it. The most widely-used application of this is in the specialised field of environmental analysis, where the daily rise and fall in outside temperature can be modelled and the effect observed. This area is dealt with in more detail in the next chapter. A more general application is simulation of moving entities, which are usually people but can also be vehicles, lifts and so on.

Simulation can model the effect of people moving around a building. The amount of use of the facilities at different times of the day will be reported and the designer can check that the provision is adequate and that people do not have to queue for long periods for the lifts, jostle each other in the corridors, crowd the rooms or overload any other facility. At the same time, the architect can check that the design is not over-lavish in the provision of facilities that will make the building more expensive to build and to run.

At present, the architect relies on experience to judge the correct amount of provision for each area. This may be his or her own experience of what has worked in the past, or it may be that of some other designer available in the form of existing buildings or in planning guides. In general, this is a good method of working, but it suffers from a couple of serious drawbacks. First, any departure from a solution that has worked before leaves the architect with no point of reference: it is impossible to be sure that the new design will work. Therefore, original solutions are discouraged and the basic designs tend to remain unchanged.

Second, because designs remain unchanged, the less obvious mistakes can become 'fossilised' and carried forward from one building to the next. For example, studies a few years ago showed that teaching rooms in universities were on average used at only one-fifth of their capacity, despite overcrowding in some rooms, and that this was largely due to an unsuitable range of room sizes[100]. These had been carried down from building to building because it had not been practicable for any designer to set up long, elaborate and costly studies into utilisation factors. So previous solutions which worked, but which were inefficient, were used. Similar errors must exist in many other types of building and can only be eliminated by systematic analysis.

Computer simulation provides a quick and cheap method of performing such analysis and allows the architect to predict how a building will perform in practice. New and adventurous designs can be tested and buildings tailored to the user, thus optimising the use of the budget. This can be especially important when designing a building such as a hospital or a university where there are often very stringent area and cost allowances. As there is so little scope for playing safe by over-provision, the designer must be certain that all facilities are adequate but not wasteful.

There is of course the danger that fitting a building too closely to its function could remove flexibility in the future when needs change. However, with greater insight into the workings of the design the architect should be able to see where bottlenecks might occur and allow for them. He should also be able to predict the effects of postulated changes in use or in the number of users of the building.

A recently-developed form of program that can be used to aid the design process is the 'expert system'. This is essentially a program that mimics a conversation with an expert in some subject or other. The user types in a question and the system may respond with an answer or perhaps a request for more information. The system uses a 'knowledge base' which corresponds to a human expert's knowledge and experience and is able to make logical inferences and deductions. It can therefore arrive at a solution in a very free manner and with the minimum of input[101].

Expert systems have only recently been applied to problems in architecture and are very much at the development stage. One promising application is in the analysis of building defects. By taking information about the type of defect, such as water penetration, timber defects, cracks in the structure and so on, and asking relevant questions, the expert systems can arrive at a list of possible causes of the fault and the probability of each.

The greatest use of expert systems at present, however, seems to be as an analytical design aid. One important aspect of design is guidance through legal complexities. There are very many rules and regulations which govern all aspects of building design, including height, distance from a boundary, obstruction of the light to other buildings and so on. The problems of obtaining planning permission are so acute in some urban areas that many architects have become very successful, not because of their design ability but because of their intimate knowledge of the regulations and their rights of appeal. This enables them to obtain permission for larger buildings than those allowed to architects who are not familiar with the technicalities.

Expert systems have been built which have been fed with the planning laws and building regulations of various cities. The designer can then 'talk' to the system and ask it about the constraints on his design. He can propose a design and have the computer tell him if it is unacceptable and if so, why. Obviously, this can be invaluable at a very basic stage of design[102].

Expert systems have also been built to aid other design problems, such as advising on the types of trees and other planting to be used in landscaping a building having regard to the soil type, the ultimate height of the planting, shading etc.[103] A system has also been written that advises on the creation of architectural working details[104]. Research is going on worldwide into systems that help with construction management, material selection, cost estimation, energy conservation etc.[105] Few of these systems have been used in practice at the time of writing, but it can be expected that in a few years' time integrated ranges of expert systems will be available for the use of architects.

Dynamic design analysis - simulation

A few years ago, there was no practical method for predicting the cumulative effect of many people acting independently. In general, problems including many elements, each behaving in an arbitrary manner, cannot be solved in a fixed number of sequential steps by applying rules and evaluating formulae. The behaviour of human beings is is normally such a problem. Finding a solution therefore entails the analysis of hundreds of separate interacting variables. This has been done by hand in the past, but it is so onerous that it was never really viable as a technique. It is not sufficient just to take a handful of variables and multiply up the result, because individual cases may have a disproportionate effect; many variables must be considered if random quirks are to be evened out. Since the advent of cheap computing, however, this problem can be solved quickly and economically. The computer can model the actions of many people each behaving in a random manner and interacting with others, and can print out the total demands on any facility over a given time span. This technique is generically known as behavioural simulation[106].

Various methods have been developed to simplify the construction of these models. Using such methods, the architect can test aspects of the design and modify it and retest until it works well enough to satisfy him. Simulation is most often used in architectural applications to predict the movement of people, but can be used in any situation involving random entities or variables, such as cars, aircraft or outbreaks of fire. These techniques have been used for lift siting, circulation analysis, canteen design, car parking provision, airport terminal layout, supermarket design and the solving of many other diverse design problems.

The answers that the architect gets from simulation are not feasible layouts, or even comments on how suitable the design is for its purpose. They are descriptions of what happens, from which the architect must decide the adequacy of the scheme. For example, if modelling supermarket checkout points, the computer will output a description of the queues that formed, giving the fluctuations in length over the day. It is then up to the designer to decide if the figures are acceptable. If the average queuing time is over five minutes this might be considered too long and the architect might try the effect of revising the model to add an extra checkout point or increasing the shop assistant's productivity by providing more mechanisation.

However, if queues are very short or form infrequently, the designer can try eliminating a point or arranging that the assistants take on extra duties such as packing the purchases or stocking the shelves. By these means the optimum provision of staff and equipment can be determined, and armed with these facts the architect can produce a layout without having to guess at figures or rely on existing solutions.

As with the critical-path technique, it is helpful to draw a diagram of the model network to give a visual impression. It is also necessary to identify individual activities and join them in sequence. The similarity ends there, however, as simulation models are typically made up of separate but interacting networks, which often have looping and feedback characteristics. This can be made clearer by constructing a model for the specific example of a single supermarket checkout point. At the checkout, which is shown in *Figure 9.1*, there are two interacting networks. The first of these is the customer network which is entered again and again by individual customers. The process of checking out might be divided roughly into the individual activities of queuing for service, waiting while the cost of the goods is added up and paying.

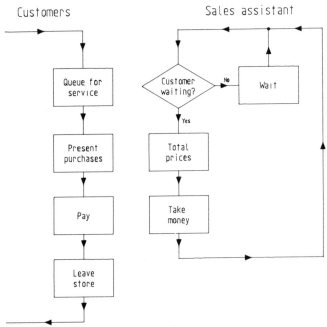

Figure 9.1 A simple simulation network

The other network describes the assistant's activities. Only one assistant is present, so this network forms a loop with the same activities being carried out for every customer. There are two basic activities: adding up the prices and taking the money, which includes giving change if necessary. When these activities have been carried out for one customer, there is a decision point, indicated on the diagram by a diamond shape. At this point the assistant checks if a customer is waiting in the queue, and if so goes through the network again. Otherwise, the decision point is entered again after a short lapse of time. This formal way of expressing the fact that the assistant is waiting for the next customer is necessitated by the nature of simulation on digital computers in which the lapse of time must be modelled in discrete intervals of, say, one second.

This model is an over-simplification, of course. The assistant might have to perform additional activities such as dispensing carrier bags, giving trading stamps or restocking the till with change. If these occupied a significant amount of time they would have to be included in the model. Activities can optionally be broken down in more detail. Thus taking the money and giving change could be separated, instead of being considered as a single transaction, and statistics collected on both tasks. The results from this might indicate that an automatic change-giving machine should be installed.

The network might also be made more informative by considering special cases. Thus if a significant proportion of customers pay by credit card or cheque, the network could divide to simulate the mix of cash and non-cash customers. The average time to fill out a credit slip or write a cheque and to verify these takes considerably longer than a cash transaction. Separating out the non-cash transactions, therefore, could yield figures that would be valuable when attempting to speed the paying process. This refined network is illustrated in *Figure 9.2*.

The model can be extended in various ways as well as refined. It could include the customers' selection time in the store and the number of items chosen, for example. This would give information on how large the body of the store needs to be and how much stock need be carried. It is up to the user to decide what information is needed and construct the model accordingly.

Inspecting the simpler model, it can be seen that there are three points at which data has to be supplied on activity rates. The first is the rate at which customers join the queue. This will obviously vary throughout the day and throughout the week. In this case there may be peaks at mid-morning and mid-afternoon, with much larger peaks on Saturdays. The model could therefore be provided with a reference table giving the intervals between customer arrivals at the queue throughout a typical weekday. These intervals are expressed as a histogram in *Figure 9.3*.

In principle, the intervals can take any value, but for simulation they must be rounded off to whole numbers of the discrete time units in which the model is constructed. This need not be a restriction as the units can be made as fine as required, although computer processing time will increase proportionately. Units of single seconds should be more than adequate in this example, and much coarser units might well give acceptably accurate results in a shorter time.

The arrival of customers is determined by such things as working hours and social conventions; therefore the histogram does not show symmetry or mathematical consistency and the only way to gather data describing it is actually to perform a survey for at least one day. This task can be considerably eased, however, if only the peak time is modelled.

Peak-time arrival can often be expressed simply as a pair of limits: customers

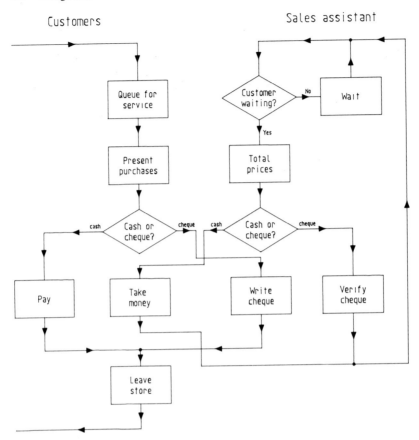

Figure 9.2 A more detailed simulation network

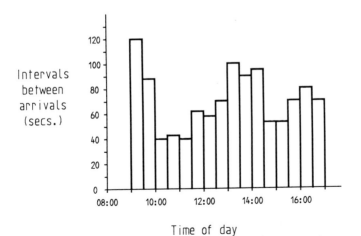

Figure 9.3 A histogram of customer arrivals

might be said to join the queue at random intervals between 40 and 60 seconds. The architect can then see how adequate the design is for the worst case.

The second activity on which information must be supplied to the model is the speed at which the assistant totals the prices. Observations must be made on the speed of a typical assistant, which will of course vary with the number of items. Because of the nature of the task, practically any time may be observed and the results are conveniently expressed as a curve from which the percentage of occurrences of a given range of times can be read off. This curve is shown in *Figure 9.4*.

As can be seen, the average time taken is about 30 seconds, the values falling off smoothly and sharply towards zero and smoothly but slowly towards infinity. This shape of curve is said to be 'positively skewed'; it indicates that most orders fall into a narrow band in the lower numbers, but a few are very quick, involving only a few items, and some are lengthy, perhaps indicating a large number of items or difficulties in assigning prices. Negatively-skewed distributions, where most

Figure 9.4 The distribution of goods totalling times

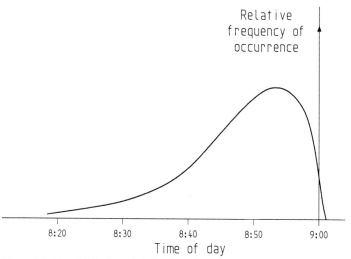

Figure 9.5 The distribution of staff arrival times

observations are recorded in the higher numbers, are much less common. One example is the time of arrival of office workers, which tends to build up towards the official time of starting work, and then falls off rapidly, most people arriving in the ten minutes before starting time. A curve illustrating this is shown in *Figure 9.5*.

The third piece of information to be supplied to the model is the length of time needed to pay. Like the previous values it is continuous in nature, but because of the type of transaction the distribution of time is symmetrical. The average is about 15 seconds and the time varies slightly either side of this, reflecting the different number of notes and coins taken and returned in change. *Figure 9.6* shows the distribution diagrammatically.

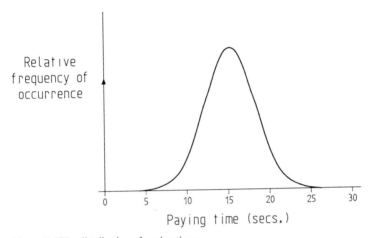

Figure 9.6 The distribution of paying times

The distributions have been shown in conventional form. Most simulation systems, however, require the curves to be presented so that, for any activity time, the percentage of times on which that time is not exceeded can be read off. This is a more suitable form for analysis, because if pencentage figures are selected at random, a suitable set of times will be provided for the simulation. This sort of curve is known technically as a 'cumulative frequency curve' and the redrawn form of the paying time distribution of *Figure 9.6* is shown in *Figure 9.7*.

Given these sets of information, the computer can simulate the activities. This model could be defined in about a dozen lines using one of the more advanced simulation languages. It is quick and cheap to model typical interactions so that many of the questions that vex the designer trying to provide for such situations can be resolved with little difficulty.

Over the years, a number of different methods have been developed for describing simulation models to a computer. The earliest method was to provide a set of sub-programs that could be invoked from a conventional high-level language, usually FORTRAN. Such facilities never became really popular because describing a simulation network, which is inherently a number of events occurring in parallel, in a language which inherently assumes events can only take place in sequence is an unnatural way of working.

A more advanced method is a high-level language developed with simulation in mind and which can therefore reflect the organisation of an activity network, but

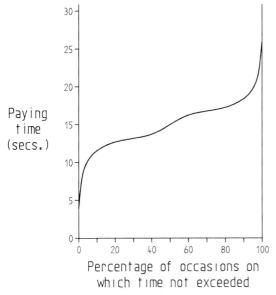

Figure 9.7 A cumulative distribution

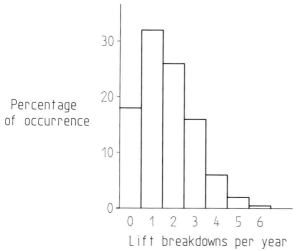

Figure 9.8 The Poisson distribution

which at a lower level has the capabilities of a conventional programming language. The advantage of this is that all the normal facilities of the computer are available.

The most recent simulation languages have cut the last ties with sequential programming. They allow the direct specification of such things as queues and the allocation of resources. Such languages are very terse in use; as an exercise I have modelled the same network in a number of different simulation systems. The results were broadly that the model description occupied 200 lines of a general-purpose language equipped with simulation sub-programs; 80 lines of a language with a simulation bias and only 27 lines of the special-purpose simulation language[107].

In the past, results were always printed out; but now that cheap screens are available it is becoming increasingly popular to show the effect of queues changing and of other results as they occur in a graphical manner. This is a very convenient method of working as the user can run through a typical day on the model in a few minutes of real time and watch how it responds to the varying demands[108].

A drawback to simulation may appear to be that a lot of time has to be spent taking observations to supply the basic data. In practice, this is not always a problem, because the data is so often reusable. For instance, in the example given previously, once the serving times have been established for a typical assistant they can be used in any other supermarket analysis and often in similar situations as well, such as totalling meal items in a self-service restaurant or counter transactions at a bank. As most architects tend to specialise in a fairly narrow range of building types, a 'library' of useful observations can be built up to cover most of the activities that need to be modelled.

It will be found in practice that many of the observations made correspond to a standard statistical distribution. The most generally applicable of these is the 'Normal' distribution that can describe the observations in hundreds of everyday occurrences. The curve is roughly bell-shaped with the observations symmetrically distributed about an average. The distribution of paying times in *Figure 9.6*, for instance, is a close approximation to the Normal distribution.

Another very common expression is the Poisson distribution. This shows how events that have an equal chance of occurring at any time or at any place will occur when observed a large number of times. For example, the number of goals scored at football matches follows a Poisson distribution. The number of flawed bricks in a load will approximate closely to an observation from a Poisson distribution as will the occurrence of fires in a large building. *Figure 9.8* gives an example of a Poisson distribution recording the incidence of lift breakdowns.

In the diagram, the curve is positively skewed, but in general the curve has no fixed shape.

A knowledge of the expected form of the observations is both a very powerful and a very dangerous aid to data collection. It can greatly reduce the number of observations that need to be taken, as the smoothed form is known, but at the same time it is easy for someone who is not a trained statistician to choose the wrong distribution and in any case in real life unexpected quirks and fluctuations often occur. Even if it seems obvious in theory to use a certain distribution it may not be valid in practice. Unless the situation is clear cut it is better to play safe and to fully observe actual events.

Over a period of time, most firms will build up or acquire ready-prepared simulation models for particular situations and certain buildings. These can then be used with relatively minor modifications on a range of different schemes.

Some organisations are marketing simulation packages to model various standard situations. The provision of lifts is a popular one and about a dozen models are available at present. The user states which lifts are to be used and the packages normally have the basic data for the popular lift manufacturers built in. Various strategies are available, such as an express lift to certain floors, or giving certain floors priority in summoning lifts. The user can then test combinations of lift type, siting and strategy and so optimise the performance[109].

The cost of lift machinery is significant in most large buildings: typically 2% of the total cost. It is therefore not difficult to make large savings in this area for relatively little effort. A byproduct is that the congestion in lift lobbies can be checked and as these typically form 2-5% of the total circulation space, important savings can be made here[110].

Other standard packages exist, including those for airport terminal design[111]; factory production lines and general circulation in buildings[112].

The HOCUS simulation system

One of the most popular and well-established simulation systems is HOCUS ('Hand Or Computer Universal Simulator'). This system was developed by P-E Information Systems of Egham, Surrey. It was first written in 1969 for mainframe computers and has undergone continuous development since. It now runs on a very wide range of computers, including the IBM PC/AT microcomputer. Over 250 organisations currently make use of HOCUS for their planning.

HOCUS enables a network showing the logical sequence of events to be described directly. At the same time, the user is able to specify the resources that will circulate within the network. In architectural applications these will usually be people, but could also be items of equipment, materials and so on. HOCUS has the great advantage of having an optional animated graphic output, by which the user can actually watch the simulation taking place.

Networks in HOCUS are made up of linked elements of two types: activities and queues. An activity is a process that takes a certain time to carry out and requires certain resources (or 'entities' as HOCUS calls them). A queue is a point in the network where resources wait until needed by an activity.

This can be made clearer by redrawing the network of the supermarket checkout point that was illustrated in *Figure 9.1* in a form compatible with HOCUS. By convention, HOCUS uses rectangles to represent activities and circles to represent queues. Both queues and activities are given unique numbers by which they are referenced and can also be given names for reasons of convenience. *Figure 9.9* shows the redrawn network.

Taking each element in turn, queue 1, named 'STORE' contains a pool of customers that will pass through the checkout from the supermarket. This queue links into activity 1, called 'ARRIVE'. The role of this element is to select customers from the store queue at the correct intervals, thus giving the effect of customers wishing to check out. Activity 1 passes the customers into queue 2, named 'WAIT' which represents the queue of customers waiting for service.

Another queue, number 3, called 'IDLE', contains the assistant when unoccupied. Both queue 2 and queue 3 link into activity 2, named 'TOTAL GOODS', which is the process of checking and totalling the purchases. It will be specified that this activity requires one customer from queue 2 and one assistant from queue 3.

When activity 2 has finished, it moves the customer to queue 4 and the assistant to queue 5 before the next activity can take place. These queues have no real existence of course, the second activity following on directly from the first, but HOCUS requires that networks be made up of alternating activities and queues. When the program is running, the customer and assistant will be in these queues for an instant of time and will then move into activity 3, named 'PAY', which is the process of paying and receiving change.

When activity 3 finishes, the customer will move into queue 6, named 'OUTSIDE', which represents the outside world, and the assistant will move back into queue 3.

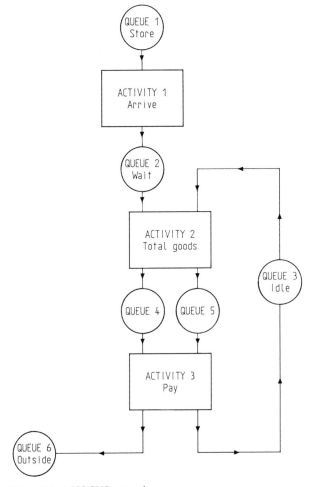

Figure 9.9 A HOCUS network

When performing the simulation, the program tests the network at each fixed instant of simulation time and moves resources through the network when possible. If an activity is not in operation at a certain time, the system will check the state of the queues that feed it. If the resources to start the activity are available, those resources will be moved into the activity element and the activity will be recorded as being in operation. At later instants in the simulation, the system will check to see if the activity has finished. When it does, the resources will be moved on into their designated output queues.

In order to be able to cope with the most complex situations in real life, HOCUS allows a range of options to be associated with elements. Activity durations may be of a fixed time, or samples may be taken from a cumulative distribution to give the effect of random individual variations.

Activities may specify in what manner they transfer resources from and to a queue. They will normally take the resource that has been waiting the longest time, but may also select a resource at random, or may select the resource that has been waiting the shortest time. This last method of 'last in, first out' is appropriate in many manufacturing situations.

Queues do not have to have real physical equivalents. For example, a simulation of car-parking might include a 'queue' of empty parking spaces which drivers select from at random.

Having designed the network, it must now be put into the system. There are two types of standard forms to prepare, one for activities and one for queues. Queues are the simplest as it is only necessary to specify that a certain queue number exists. Optionally, it can be given a name and also a limit to its size. Thus the queue information for the checkout point simulation would be pairs of queue numbers and queue names as follows:-

```
1   STORE
2   WAIT
3   IDLE
4   DUMMY
5   DUMMY
6   OUTSIDE
```

The activity forms are more complex. The first line is the activity number and optional name, and also the number of resources the activity uses. The next lines list the resources and specify how they are moved from and to the queues. The last line details how long the activity will take. Thus the data for activity 1 is:-

```
1   ARRIVE 1
E1  CUSTOMER 2T
50
```

The first line states that this is activity number 1, named 'ARRIVE' which uses one resource. The second line defines this resource. Its name is 'CUSTOMER' and it will be taken at random from queue number 1 and put into the tail of queue number 2. The third line specifies the intervals at which this will happen. In this simulation of peak time arrivals, the activity duration is fixed at 50 seconds, so the effect is that every 50 seconds a customer is passed through the activity into the queue for service.

The data for the next activity takes the form:-

```
2   TOTAL GOODS 2
H1  CUSTOMER 3T
H2  ASSISTNT 4T
F1
```

The first line states that this is activity number 2, named 'TOTAL GOODS' and uses two resources. The next line states that the resource 'CUSTOMER' will be taken from the head of queue number 2 and put into the tail of queue number 4. The third line states that the resource 'ASSISTNT' will be taken from the head of queue number 3 and put into the tail of queue number 5. The last line details that the time taken will be taken at intervals specified by distribution F1 which is supplied separately. The distribution of goods totalling time was illustrated in *Figure 9.4*. It must be converted into cumulative distribution form and the data extracted.

The data for activity 3 is very similar and has the form:-

```
3    PAY 2
H3   CUSTOMER 5T
H4   ASSISTNT 2T
F2
```

This activity uses the customer and assistant resources from the appropriate queues, but the time taken is specified to be selected from distribution F2, which was illustrated in *Figure 9.6*.

Thus the entire simulation can be represented as a network of nine elements and described fully in about 17 lines of data.

After typing the data into the system, the network will be drawn on screen. It is therefore very easy to identify mistakes in the information prepared. The user can then ask for the simulation to be run. This can be done in two ways. Most vividly the screen can show an animated representation of the system. Each resource will be identifiable and they will move about the network. The user can therefore actually watch as queues build up and empty and as people or facilities are occupied or stand idle. The animated network can be a schematic representation for normal working, but facilities are provided to paint the screen as required with full-colour images of equipment or moving entities and this latter form can be very impressive in public relations exercises or formal presentations. *Figure 2.4* showed an animated version of a simulation of a maternity ward in a hospital. This illustration is taken from a real-life exercise which was carried out to check if the number of delivery rooms provided was correct.

Alternatively, the user can instruct the simulation to move forward in time a certain period and then display the static situation at that time. This is of course useful for stepping through a process an hour or a day at a time.

It is often necessary to print out the statistics on activity use and queue behaviour for consideration. A wide variety of outputs is possible, but an overall view can be obtained from reports that the user may especially construct. One useful form is the Summary Performance Report, which amongst other things can give the proportion of time an activity is in use and the sizes of the queues. *Figure 9.10* gives a tailored Summary Performance Report after running the checkout simulation for a day.

```
                    SUMMARY PERFORMANCE REPORT
                    **************************

Assistant occupied                                    95.6 %

Average number of customers waiting                    1.2
Maximum number of customers waiting                    6

Average time spent waiting                            59.0 seconds
```

Figure 9.10 A summary performance report (courtesy P-E Information Systems)

It can be seen that the assistant was occupied for over 95% of the time, but this is an expected result at peak periods. The queue statistics are more informative: the longest queue consisted of 6 customers and the average queue length was 1.2 customers. The average time spent in the queue was 59 seconds. The architect might conclude from this that the queue times are short enough for a single checkout point to be satisfactory.

Expert systems

With the use of expert systems, we begin to move into the field of 'artificial intelligence', in which the computer attempts to mimic the flexibility and deductive powers of the human mind. Conventional computer programs rarely volunteer information unasked. They will answer direct questions, but if those questions are misguided because of the user's lack of knowledge, then the results will probably be totally irrelevant. A human being does not operate in this manner; if another person should ask questions that are clearly based on false assumptions, then the questioner will be corrected. Also, a conventional computer program is very rarely able to justify any answers it gives. It evaluates the formulae and produces the results, but it is incapable of giving the reasoning behind its actions. A human expert on the other hand is always expected to be able to justify his decisions.

Essentially, an expert system mimics a human expert in a specialised field. The expert will have a wide knowledge of his subject, will suggest courses of action as well as commenting on proposed actions and will be able to say why a conclusion has been reached. The following example shows a very simple use of an expert system to design stairs. The user input is in italics.

Hello. I can design stairs to meet British Standard No. 5395.
Is this a public, semi-public or private building?
Public.
How many floors are you considering?
1.
What is the floor-to-floor height in millimetres?
2400.
The optimum arrangement is 16 steps having 150 mm rise and 300 mm going.
What is the minimum arrangement?
The minimum arrangement is 14 steps having 171.4 mm rise and 280 mm going.
Why?
The standard states that the minimum going in public buildings is 280 mm.
The standard states that the maximum steepness is when $2 \times$ rise $+$ going $= 700$.
The standard states that the maximum rise in public buildings is 180 mm.
Set the rise to 160 mm.
The new arrangement is 15 steps having 160 mm rise and 280 mm going.

Most expert systems consist of four parts: a knowledge base, a situation model, a user interface and a controller. The knowledge base is equivalent to the knowledge that a human expert would possess in his field. The situation model is created every time the program is executed and represents the data and conclusions that are built up on these days use computers, it is quite possible that the data shape can be transferred almost automatically. Otherwise, if a digitising available, it does not take long to input spot heights from a drawn plan. the spot heights, terrain modelling programs attempt to interpolate values the known heights. Some have better methods of doing this than others, spot heights are not too far apart the estimates are likely to be good. the model in the computer, various outputs can be obtained. A contour be drawn as can any cross-section through the site or a profile of the site point. The architect often needs to reshape the site in order that building mence. Earth may have to be banked up in some parts and cut away in process known as 'cut and fill'. Terrain modelling programs can measure

'If site entrance less than 2 metres, then heavy plant cannot enter.'

This could be an entry in an expert system for site management. The situation model would be interrogated for the width of the entrance, then the conclusion on access for heavy plant would be added to the model. A more complex rule in the same system might be:-

'If cement available and sand available, then mortar can be mixed.'

In this case the situation model would be checked for the availability of both cement and sand and the conclusion on mortar added. In practice of course, more than one conclusion may result from a set of fulfilled premises.

The words 'and' and 'or' have very specific meanings in computer usage. If two premises are linked by the 'and' conjunction then both the premises must be true for the conclusion to be true. If linked by 'or' then either or both must be fulfilled for the conclusion to be true.

These sort of rules where conclusions are either true or false are rather over-rigorous in real life. A human expert is accustomed to work with incomplete data and still arrive at conclusions. This is because the human is able to decide (perhaps intuitively) which of several cases is the more likely and so build up a list of solutions, some of which are more probable than others. The more advanced rule-based systems use the concept of 'if (premise) then (probability level) likely that (conclusion)'. Some rules from a simple building fault diagnosis system will illustrate this more clearly:-

'If cracks present, then it is 40% likely that vibration is present.'
'If cracks present, then it is 60% likely that subsidence is present.'
'If heavy traffic nearby, then it is 70% likely that vibration is present.'
'If clay soil and dry summer, then it is 20% likely that subsidence is present.'

If the user gives the basic information that cracks are present, the system builds up a set of two conclusions in the situation model, one set of which is more likely than the other. The system will then ask questions about traffic levels and soil type to try to refine its guesses. At the end of the run it will print a list of alternatives with their probabilities.

Given a set of rules in the knowledge base, the controller may decide to access them in different ways. The methods described above assume that the data for the premise is given and the conclusion drawn. This access method is called 'data-driven'. Many systems, however, access the rules in reverse so that information on observed conclusions allows the system to deduce what preconditions were necessary. This is called 'goal-driven' access. The principle can be made clearer by using an example from a site management expert system. The system will typically make provision for calling site meetings from time to time. The rules as to when a meeting can be called might be:-

'If a chairman is available and more than 5 senior site staff are available, then a meeting may be held.'
'If the architect is available or the Clerk of Works is available, then a chairman is available.'
'If the architect has no prior engagements, then the architect is available.'
'If the Clerk of Works has no prior engagements, then the Clerk of Works is available.'

A goal-driven system starts from the goal it hopes to achieve, that of calling a meeting, and checks if the necessary conditions regarding the chairman and site staff are in the situation model. If they are not, it tries to obtain information on these conditions by regarding each condition as a 'sub-goal' on the way to the ultimate goal. Beginning with the chairman availability sub-goal it checks the other rules to see which conditions must be present to achieve that sub-goal. A 'tree' of alternatives is therefore built up. At the end of the process the system would expect to find the architect's and Clerk of Works' appointment diaries in the situation model. If it did not it would ask the user for information.

Goal-driven access systems tend to be more efficient than data-driven systems. They are also especially suited to certain types of problem. For example, in medicine the doctor is aware of the patient's symptoms (the 'conclusions') and is trying to diagnose the disease (the 'premise').

Rule-based systems are the most popular and useful in practice. However, there are other ways in which to organise the knowledge base. One such method is the use of 'frame structures'. This makes use of the concept of familiar situations, or 'frames', in which a human being has a good idea of what to expect. Such systems have application in areas of artificial intelligence such as speech recognition. An example might be given of an architect discussing a building with a colleague on a poor telephone line. His colleague might use a word that sounded like either 'beam' or 'bean'. The architect would guess with high probability that the first word was the correct one because of the context in which the discussion was taking place. Frame-structured expert systems use similar techniques and so are much less likely to reach the bizarre conclusions computer programs are notorious for.

The user interface varies widely from system to system. Many allow only a fixed number of alternatives to a question; often for instance only 'yes' or 'no' will be permitted. Some, however, use complex techniques that allow the user to express himself naturally. So the user could type 'Yes', 'OK' or 'That's fine' and be understood, or could type a complex sentence such as, 'We have cement but no sand' and have the computer extract the relevant facts. It is not true to say that the best systems all have advanced natural language user interfaces; because of the inherent scope for ambiguity and the volume of human language, many software designers prefer the terse form.

Expert systems are still in the development stage and few immediately useful packages for the architect are available. This situation can be expected to improve shortly. One of the reasons for expecting rapid improvement is that several suppliers now offer expert system 'shells'. These are systems that have no fixed knowledge base, but provide convenient means for constructing one. It is therefore a relatively simple task to build an expert system in virtually any subject[113].

Design generation - plan layout

The most popular type of design-generation program is that which produces plan layouts. Very many such programs have been written, although few are now used in practice.

The basis of virtually every generative design program is a matrix of the interactions between rooms or functional areas. Typically the user will decide on a list of the rooms that will make up the building and then will assign each pair of

should be close together, as with an operating theatre and the surgeon's scrub-up room in a hospital; the value 0 that there is no connection between the rooms and so relative positioning is unimportant and the value -5 that the rooms should be well separated, as with a kitchen and a bathroom, for instance.

As pairs of rooms are involved, the matrix is symmetric and can be conveniently shown as a half-matrix. *Figure 9.12* shows a typical interaction matrix for a group of rooms in a hospital.

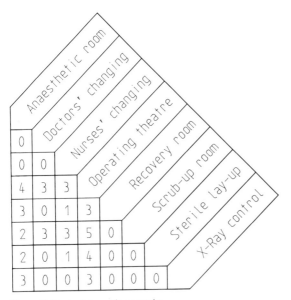

Figure 9.12 An interaction matrix

An objective way to generate an interaction matrix might be to study an existing building and total the number of journeys between each pair of rooms. A program using this data could then minimise circulation costs in the design, which might be an important factor. The drawbacks are that this sort of data is laborious to collect and that it does not allow for necessary separation between rooms.

Separation might be required for hygenic reasons or to cut down on noise interference or even for purely psychological reasons. I was once involved in using a program to minimise the circulation in a hospital design and one of the results located the mortuary close to the geriatric ward as there was quite a lot of traffic between them. The architect felt, however, that the proximity might not be good for the patients' morale!

Given an interaction matrix on some basis, programs are available to generate a design at one of two levels. The first is a schematic design in which room symbols are laid out in a manner that corresponds to the optimum arrangement if the considerations of room size and building costs are ignored. This form of output corresponds to a designer's early 'bubble diagrams' in which he scribbles different arrangements in order to clarify within his own mind the relationships involved. *Figure 9.13* shows such an output that was produced by the MAGIC program which is supported by the ABACUS group.

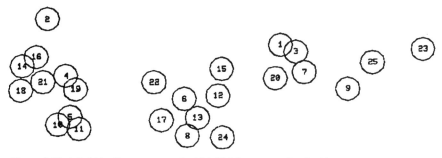

Figure 9.13 A bubble diagram output by MAGIC (courtesy ABACUS)

The second level of output takes the process a step further and uses the required room areas and some simple rules of building to produce an outline plan.

For either type of output, the first essential is to reduce the interaction matrix into a form that can be regarded as separation distances in two or three dimensions. This must require some adjustment of the values in the matrix because, for example, a group of five rooms might need to be close together, while a sixth room needs to be close to one member of the group and distant from another.

In the 1960s, many methods were devised to perform this transformation with as little distortion as possible. One of the most straightforward methods was the association-graph technique[114]. This method can be thought of as drawing a diagram consisting of scattered points, each point representing a room. A line is drawn between the pair of rooms with the strongest interaction. A link is then drawn for the next strongest interaction, and so on. The location of the points is adjusted where necessary to prevent links cutting across earlier links, but if it is impossible to prevent crossings, the weaker link is ignored. The end result is that the points are forced into an arrangement taken to represent the optimum positioning of the rooms.

Additive techniques position the room which has the strongest overall interaction at the centre of the plan, then place adjacent to that room the room that is linked most strongly to it, then position the room most strongly linked to that pair, and so on. At each stage the next room added to the plan is the one that has the strongest links with the group already positioned and is located relative to the strength of its links with each group member[115].

Permutation techniques start with a proposed layout and test all possible exchanges of position between every pair of rooms. The exchange that most decreases the total 'cost' of the layout, as measured by separation distance multiplied by interaction matrix values, is then effected. This technique has been extended to triads of rooms and to allow for some rooms to have fixed positions[116].

Among the more mathematically-based techniques is multi-dimensional scaling in which the values in the matrix are regarded as distances in n-dimensional space. One dimension is removed at a time, with the distances being adjusted as little as possible to accommodate it, until finally the matrix represents an arrangement in two or three dimensions[117].

Another mathematical technique is based on 'linear programming' methods, which are widely used in the business field to optimise deliveries to customers. This technique requires the user to assign a cost to each room being at each of a number of possible locations on a plan. This cost is typically based on the amount of traffic to the room and the distance between a proposed location and the centre of the plan. Given this information, linear programming methods can be used to find the arrangement that gives the lowest total cost[118].

Cluster-analysis methods attempt to distinguish groups or 'clusters' of rooms that hang together strongly according to certain criteria. There are many different possible criteria but a typical one is that every member of the cluster must link to every other member with at least a minimum interaction. By relaxing the criteria, some of these groups will coalesce into larger groups until finally all activities are contained within one group which represents the complete building[119]. A cluster analysis applied to an architectural firm is shown diagrammatically in *Figure 9.14.*

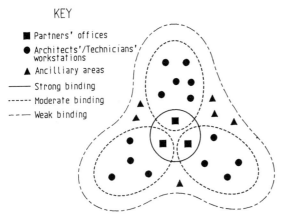

KEY

■ Partners' offices
● Architects'/Technicians' workstations
▲ Ancilliary areas
—— Strong binding
----- Moderate binding
――― Weak binding

Figure 9.14 A cluster analysis diagram

The partners' offices tend to cluster together because of frequent important discussions. The design teams also form three separate strong clusters. At the next level, each partner clusters with the design group he is responsible for. These clusters therefore overlap with the stronger one, which is a common occurrence in real life, although not all cluster-analysis programs allow overlapping because it makes evaluation much more difficult. At the lowest level of binding, the administrative areas such as the secretaries' room, the accounts department, the conference room and the library coalesce to form a complete group.

These are some of the most popular techniques, although many others have been developed which are of greater or lesser validity.

To go on to produce a layout plan, a program needs to know the required area of each room. Simple rules can then be applied on the acceptable room proportions, on which rooms require an outside wall, and on structural limitations, and then some sort of plan can be drawn. The more elaborate programs of this type allow the user to specify several storeys and take into account the greater difficulty of vertical as opposed to horizontal communication. As well as trying to minimise circulation, many programs also attempt to optimise construction costs by reducing external wall lengths or avoiding awkward plan shapes.

Most layout-generation programs do not require or allow user interaction, but produce results completely automatically. Programs that do allow some interaction have the advantage that the designer can override some of the program's decisions on the grounds of other criteria.

Design-generation programs have essentially failed because no matter how ingenious the mathematical techniques applied, the basis is that minimising communication is of great or over-riding importance, and this is not a valid assumption except in special circumstances such as factory production lines. In most buildings, communications are not a problem and the architect will give much greater priority to an interesting or aesthetically pleasing layout and to other

```
WELL WARREN HERE'S MY DESIGN ON YOUR INFORMATION
```

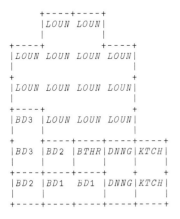

```
           +----+----+
           |LOUN LOUN|
           |         |
+----+                +----+
|LOUN LOUN LOUN LOUN|
|                   |
+                   +
|LOUN LOUN LOUN LOUN|
|                   |
+----+              +
|BD3  |LOUN LOUN LOUN|
|     |              |
+     +----+----+----+----+
|BD3  |BD2  |BTHR|DNNG|KTCH|
|     |     |    |    |    |
+----+----+----+    +    +
|BD2  |BD1  BD1 |DNNG|KTCH|
|     |         |    |    |
+----+----+----+----+----+
```

```
              ROOMDATA
              --------
ROOM      ROOM     ROOM AREA
 NO       NAME     SQ FEET
     1  BED1        128.00
     2  BED2        128.00
     3  BED3        128.00
     4  LOUNGE      832.00
     5  DINING      128.00
     6  KITCHEN     128.00
     7  BATHROOM     64.00
TOTAL AREA         1536.00
```

```
THE FOLLOWING SYMBOLS WILL BE USED FOR FUTURE PRINTOUTS
     1  BED1      ⊔
     2  BED2      ○
     3  BED3      ○
     4  LOUNGE    ⊘
     5  DINING    φ
     6  KITCHEN   ▽
     7  BATHROOM  △
```

```
LENGTH OF PERIMETER        176.00    FEET
LENGTH OF INTERNAL WALLS   120.00    FEET
LENGTH OF MODULE             8.00    FEET
FIGURE OF MERIT (COST)     246.00
```

```
WELL WARREN, WHAT DO YOU WANT TO DO NOW?
CHANGE THE AREA OF THE LOUNGE
```

Figure 9.15 An output from a plan-generation program (courtesy University of Sydney Dept. of Architectural Science)

subjective factors that will very much concern the occupants but which cannot be expressed in an objective manner acceptable to the computer.

Communications have such a low priority in fact, that some architects have deliberately designed office blocks and other buildings where the work tends to be sedentary in such a way that the staff will have to walk further during the day and so get more exercise. Similar policies have been followed in some scientific and research establishments to make the staff mix more and so encourage conversations that could lead to the exchange of valuable information.

The University of Sydney plan-generation program

One of the most advanced plan-generation programs was written at the Department of Architectural Science of the University of Sydney in 1973[120]. Despite its age, it represents a peak in the development of this application. It was also ahead of its time in that the programming team went to some lengths to make it 'user friendly' and to build in natural language facilities.

The program is extremely 'chatty' so as to give the impression of carrying on a dialogue with the user. It addresses the user by his or her Christian name in such terms as: 'Well, John, here's my design,' and encourages him frequently by commenting on the responses with words like: 'Good!', 'Great!' and 'Bonza!'

More importantly, perhaps, it has the ability to extract the significant words from a sentence like: 'I want to add some rooms', where most programs would require explicit commands. The programs will respond to requests for help at any point and will list the options available to the user or explain in detail what input is required.

As input the user is expected to provide a list of room names and areas and a table of the interactions between the rooms. The program prompts for these by requests like: 'Please type in your room names'.

As output the program will produce a plan on the terminal; a typical example is shown in *Figure 9.15*.

The user can then elect to modify interactions and change room areas. He can also specify which rooms have a window, and has some control over the positioning of rooms. After each set of changes the program types the lengths of the external and internal walls in the revised design and gives a cost factor which shows the losses in economic terms caused by the interventions.

Chapter 10
Environmental analysis

The need for environmental analysis

In recent years, analysis of the environment in a building has been turned over almost entirely to the mechanical and electrical services engineer. The architect has normally relied on experience and rules of thumb to produce an adequate basic design and the engineer has worked out the details of the lighting fittings and the heating and ventilation plant needed to service the design. Occasionally, the engineer would advise the architect to change the design in some respect if it would give significant advantages in simplicity or cost savings.

In a period of cheap energy, this policy was normally sufficient. The architect might quite reasonably be generous in the specification of window sizes, for example, and a larger heating plant could be provided to compensate for the extra heat losses; the costs of constructing and running the building would not be greatly influenced. Thus when energy saving was not an important requirement of the design, the weak liaison with the engineer did not particularly matter.

However, since the 1973-74 oil crisis, priorities have had to change radically. Energy in all its forms is now much more expensive and seems likely to become even more expensive with time. There is now great pressure on the architect to design buildings that require less energy and therefore cost less to run, and this may strongly influence the basic design of the building. The building may become more compact, have a different shape, a different orientation, or less glazing. In this new situation, close liason with some source of environmental analysis is vital; the current method of sending the design to and fro between the engineer and the architect for analysis and refinement is clumsy and inefficient. It may lead to a poor compromise, or to the engineer having undue influence over the design.

The problem has been made more acute in some countries by energy-saving legislation that affects building design. For example, in the UK the Building Regulations since 1978[121] require non-domestic buildings to meet strict insulation criteria. Legislation also exists in most developed countries to enforce environmental standards. In any case, user expectations of comfort have been steadily rising and most people expect their workplace to be adequately heated and ventilated during both summer and winter; they also expect reasonable standards of acoustic insulation and privacy. If a building is not to be very expensive to run, therefore, careful analysis of its environmental performance is essential.

A solution to the problem may well be provided by the use of the computer. Most of the analyses required for environmental design are straightforward and objective, if mathematically complex. The formulae and standards required have been the subject of much research over the years and are widely published by such bodies as the UK Chartered Institution of Building Services Engineers[122] and the American Society of Heating, Refrigeration and Air-Conditioning Engineers[123]. These guides provide a reliable basis for calculation which can readily be given to a computer. The architect can therefore evaluate the building and, provided he knows the basic principles involved, can alter it until it satisfies the requirements. The computer can be used to check on the adequacy of daylighting, specify the amount of artificial lighting required, give the size of the heating plant taking into account all gains and losses, and even print out a comparison between different fuels for the heating plant in terms of installation costs and running costs. If the architect has access to a microcomputer, one of the many packages available can be used to test environmental factors and integrate them with the other principles and concepts of the design.

Program structure

Environmental analysis is a computer application which, as with, for example, visualisation or critical-path analysis, needs so much data to be specified that there is often a danger of its not being cost-effective. In itself, the data is easy to collect as most of it consists of simple parameters such as structural dimensions and U-values which can be measured or taken from standard reference tables[124]. Despite its simplicity, there are so many items of information to be supplied that data collection tends to be a long and tedious business.

In the past, programs tended to deal with just one aspect of environmental performance. Daylighting was a popular choice and such subjects as solar gain and noise transmission were also tackled. Such programs still exist, but the current trend is towards programs that analyse a wide range of factors at the same time, or sets of programs that analyse different factors but use the same basic set of data. This is possible and convenient because a lot of the data values are common to many analyses. Room dimensions particularly, are required in almost all calculations. Also, most of the environmental factors influence each other. For example, the level of natural light governs the amount of artificial lighting required during the day and both the area of glazing and the heat gain from lighting fixtures affect the thermal balance.

Programs that analyse single factors may be slightly faster and give an extra place of accuracy than manual methods. For instance, a survey into daylighting analysis programs indicated that they are a little faster than manual methods such as daylight protractor and Waldram diagrams; but the process of accessing the computer program and typing in the information will make traditional methods more popular with most designers[125]. However, for a relatively small extra effort in data preparation, the complete environmental performance can be evaluated and this will be dramatically more cost-effective.

At an even more inclusive level, there exist integrated computer-aided design systems which require a three-dimensional model of the building to be defined. These are usually most concerned with producing drawings, but commonly offer environmental analysis as well. This approach is obviously very simple and cost-effective as the environmental analysis is offered in addition to the other results and for very little extra effort. There are a couple of drawbacks, however.

The first is that this approach cannot be done piecemeal: the whole manner of working on the job has to be changed from manual methods to almost completely computer-orientated techniques, and this might be considered unacceptable. The other drawback is that the results can often only be obtained when the building has been designed in some detail and so a lot of work will already have been done in the design, which may be wasted if its performance is inadequate. It may, for instance, be necessary to have sized and positioned all the windows before an accurate assessment of heat loss can be made. In contrast, programs orientated towards this type of calculation may allow data to be input at a much more general level. Glazing could be expressed as a percentage of the surface area, for instance. The latter approach will be much more preferable at the sketch design stage.

Lighting analysis

Of the many aspects that go to make up the environment, daylighting is perhaps the most basic and the most complex. It is complex not because the formulae to evaluate it are particularly difficult but because the presence of windows has psychological and aesthetic implications that cannot be evaluated objectively.

Given a window description and other data on, say, external obstructions, a computer can easily calculate the daylight factors at any point in the room; these can then be checked for adequacy and are sometimes the subject of legislation. The size of the window will also affect the amount of heat loss and solar gain experienced; thus the architect may wish to reduce the window sizes and rely on permanent supplementary artificial lighting, retaining the windows largely for the psychological effect they have on the occupants. The computer can calculate and put a price on the savings made by this and similar decisions.

Figure 10.1 shows daylight factor contours in a room. This output was produced by the IDLE program which is distributed by Facet Ltd. From this sort of result it is easy to see dimly-lit areas or excessive variation in illuminance.

The calculation of artificial lighting levels is much more straightforward as the subjective elements are less marked. Given the daylighting levels, the computer can calculate the level of artificial light required for any time of the day or year according to the published standards[126]. If the architect specifies a particular lighting fitting and defines it, the computer can often go on to calculate the number required, the total voltage they will consume and the amount of glare they produce at various points in the room. This last analysis can be very important: it is easy to specify a few powerful fittings that will give the correct illuminance level, but glare may well make them uncomfortable to the occupants of the room.

Thermal analysis

The most important aspect of the environmental performance of a building from the financial point of view is the heating or cooling necessary to maintain comfort. Again, recommended standards exist against which the performance can be measured; thus evaluation is straightforward although complex.

The complexity arises because thermal analysis cannot be performed adequately by non-dynamic analysis; that is by the simple evaluation of formulae. Because the outside temperature will vary throughout the day or night and because the building

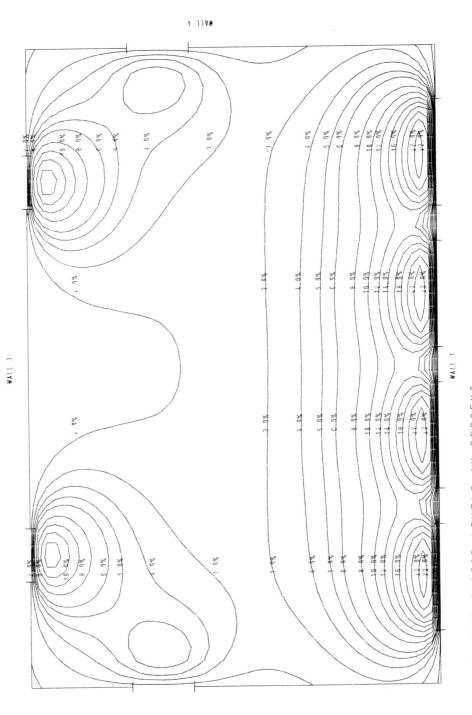

WALL 4

WALL 3

WALL 1

WALL 2

DAYLIGHT FACTOR LEVELS IN PERCENT
SCALE OF THIS PLOT 1 TO 200

structure has a thermal 'lag' as it absorbs and gives off heat, the problem is one that requires dynamic analysis and the results will be a graph rather than a simple reading[127].

Given the dimensions and the structural materials used, the computer can calculate heat losses through the fabric of the building. Losses will also occur through ventilation and data on this should also be supplied for exact results. There will be heat gains from the occupants, from machinery if present, from the lighting fittings and from solar gain. This last will of course vary not only with the season and time of day but also with the building's position and aspect and with the amount of shading from surrounding buildings and from its own shape. Significant thermal gains will also be made from the heat exchangers on the air-conditioning plant as they recover heat from the expelled air. The thermal lag of the structure and its contents will also be significant as it takes in heat during the day and gives it up when the heating is switched off at night.

Because of all the variables, the calculation of the thermal balance of a building involves long calculations, but as the required standards are clearly set out there is little scope for subjective assessments; the subject is therefore ideally suited to computer methods. A typical program can perform all the calculations in a few seconds and specify the plant required to deal with the largest heat losses and gains that would be encountered[128, 129]. If the architect decided from this summary that the plant was too large and expensive, a breakdown could be asked for giving the heat losses through each surface. From these results, the most promising options should be clear; perhaps more insulation for the walls, a smaller glazed area, double glazing, a different orientation, or more radically, a different shape for the building[130].

Most usefully, analysis can be performed at various levels of detail. Thus the designer can define the building envelope only and find the performance for the whole shape; he can then try modifying this and at the end of the sketch design stage will have a fairly precise knowledge of how much it will cost to run. At the outline-design stage, individual rooms with varying needs and conditions can be defined and optimised at this level. This method of submission and evaluation can give the architect much greater insight into the environmental implications of the design than is usually available.

Programs also exist to evaluate the trade-off between the capital cost of different heating systems and their running costs. For example, electric heating is cheaper to install than gas heating but costs more to run. From the figures provided, and having regard to the expected life of the plant and prevailing interest rates, the architect can decide which fuel would suit the building best[131].

One aspect of building design which can conveniently be associated with thermal analysis is condensation checks. With modern methods of heating and new methods of construction, condensation can occur unexpectedly within the building fabric and can cause serious damage. *Figure 10.2* shows an output from the ICE (Insulation and Condensation Evaluator) program distributed by Ibbotsons Design Software. The potential condensation risk in the outer brick leaf can be seen immediately and modification to the design made.

Sunlight analysis

Solar gain is one of the factors in thermal analysis. The internal heat gain can arise in two ways. The sun's rays may fall on external surfaces and appear as a heat gain internally at some later time, depending on the thermal lag of the fabric and on the difference between the internal and external temperatures. Alternatively, the rays may fall directly on internal surfaces through unshaded windows. This latter aspect can usefully be considered as an analysis in its own right, because sunlight has a psychological effect on the occupants of a room quite apart from the temperature change. The analysis can also be valuable in climates where solar heat gain is a major factor.

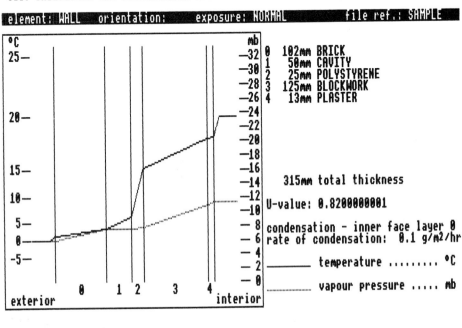

Figure 10.2 Condensation risk in a multi-layered structure (courtesy Ibbotsons Design Software)

Programs are available that will calculate the parts of a building that are shaded at a particular time of day and day of year for the appropriate latitude[132]. The shading may come from surrounding buildings or from the building's own facades, including window recesses. From the results the designer can see which rooms receive direct sunlight and at what times. This could be important for different reasons: in a temperate climate, an architect might be concerned that every room should receive some direct sunlight on each day of the year, while in Mediterranean or tropical climates the aim might be to shield the windows from as much direct sunlight as possible.

Figure 10.3 shows an output from the SHADOWPACK system which was developed by R J Peckham of the Commission of the European Communities Joint Research Centre. The system is a flexible one which can calculate and draw shadows both externally and within rooms and plot contours of their intensity.

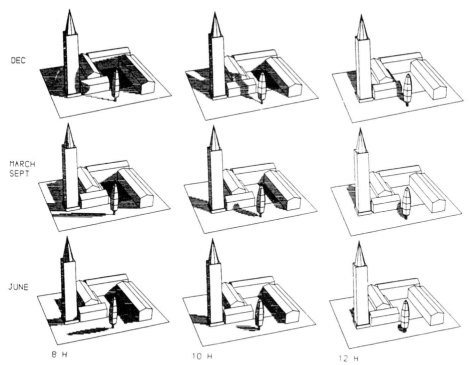

DEC

MARCH
SEPT

JUNE

8 H   10 H   12 H

Figure 10.3 Shadowing variation in a building complex (courtesy the Commission of the European Communities Joint Research Centre)

Air-conditioning analysis

Air conditioning of a building costs much more than simply heating it. There are also extra design problems because of the need to provide bulky ducting and to position vents to provide the requisite number of air changes without creating draughts.

Computer programs exist that can calculate the heating and cooling loads on an air-conditioning plant in a similar manner to those that calculate heating plant capacity, but with the additional factors of carrying conditioned air around the building and recovering heat from stale air. Again, the many sources of heat gain and loss with their different cycles according to the time of day and thermal lag can be consolidated into a total loading cycle given a desired internal temperature and a range of external temperatures. In addition, many air-conditioning programs can size pipe and duct networks, giving a compromise between wastefully large ducts with low air speeds and narrow ducts with high air speeds that take a lot of energy to drive. In large modern buildings, service ducts take up a sizeable proportion of the total volume, and so a quick and accurate method of calculating minimum duct sizes will often result in a useful gain.

More elaborate programs exist that can calculate air movement patterns throughout a building. A study of the output from these will give the designer insight into the movement and speeds of the air currents so that vents can be positioned where they will be most effective, and draughts and pockets of stagnant air can be eliminated. In some buildings an understanding of air movement patterns is vital to an adequate design. In hospitals, for example, sterile areas should have higher air pressure than their surroundings to prevent cross-infection from airborne viruses. Similarly, a kitchen area should have lower air pressure to prevent the escape of steam and cooking smells. Smoke control during fires also depends on correct air pressure. If given data on the ducting network, computer programs exist which can predict the spread of smoke and flame during a fire[133]. In buildings such as theatres and cinemas, such information can be vital to a safe design.

Figure 10.4 shows an output produced by a computer program used at the Fire Research Station in the UK. The spread of smoke from the source of a fire can be seen very vividly.

Acoustical analysis

The last important aspect of environmental performance is the acoustical behaviour of the spaces. There are several different considerations, acoustic isolation being perhaps the most important in most cases. Normally this is the amount of interference with speech from outside sources or from other sources within the space. Noise-criteria curves have been published that lay down standards for acoustic isolation by giving the maximum permissible background noise over the range of frequencies for different activities. Several computer programs exist that take data for external noise sources, calculate the way they will be transformed by passing through the building fabric, integrate them with internal noise sources from conversations and from business machines such as typewriters and then match the result against the noise-criteria curves. These programs often contain standard spectra for certain commonly-met noise sources such as traffic noise and aircraft noise to simplify data preparation.

With the current popularity of open-plan offices, the considerations of aural privacy have become very important. If the office is to operate efficiently and comfortably there must be a minimum of interference between conversations and telephone use at different points, and the annoyance from typewriters and other machines must not be excessive. With computer analysis, the architect can check the number of people that can reasonably expect to work in a certain space and can check the effectiveness of screening and absorbent surfaces[134].

Computer programs have been written to test other specific acoustic aspects of a design. The noise level in ducted air-conditioning systems is a popular application area and often forms part of integrated packages. Also, the location of fire alarms for best audibility is subject to guidelines and can be calculated by computer[135].

When designing rooms for people to listen to speech or music the consideration of reverberation time becomes important. A balance must be struck between sound clarity, which implies a short reverberation time, and sound reinforcement, which involves longer reverberation times. The ideal requirement unfortunately varies between speech, which needs clarity most of all, and music which needs reinforcement varying with the type of music performed: organ music needs more than chamber music for instance. Again, quite simple computer programs can calculate reverberation times given the room dimensions and finishes on the surfaces. Absorption coefficients for different materials can be held in the computer and this will simplify data preparation. The output will give a reverberation time, and if it is not satisfactory the surfaces' finishes can be altered.

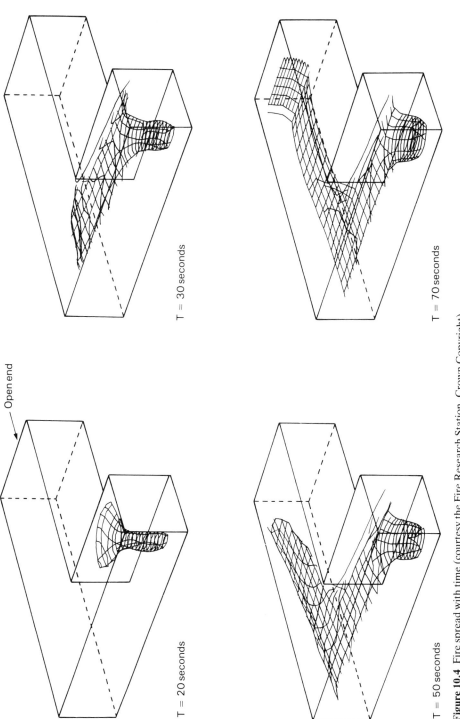

T = 20 seconds

T = 30 seconds

T = 50 seconds

T = 70 seconds

Open end

Figure 10.4 Fire spread with time (courtesy the Fire Research Station, Crown Copyright)

At its most advanced level, acoustic analysis can trace the direction and strength of the sound 'rays' coming from a stage and arriving at the listeners' ears. Because of its complexity, very little work has been done in this field, but the firm of Artec Consultants Inc in New York has developed a program called IMAGES[136] which traces the rays as they bounce about a room becoming weaker with distance and the absorbency of the sufaces they meet. The ray paths are calculated by an ingenious technique which will find all geometric paths. *Figure 10.5* is an output from IMAGES in which the listening point is on stage.

The program has been invaluable in predicting acoustic deficiencies such as echoes and patches of weak sound. One of the most important findings coming from the study of its output is that fan-shaped auditoria, which have been a popular choice for many years, are acoustically very much inferior to those that are rectangular in plan.

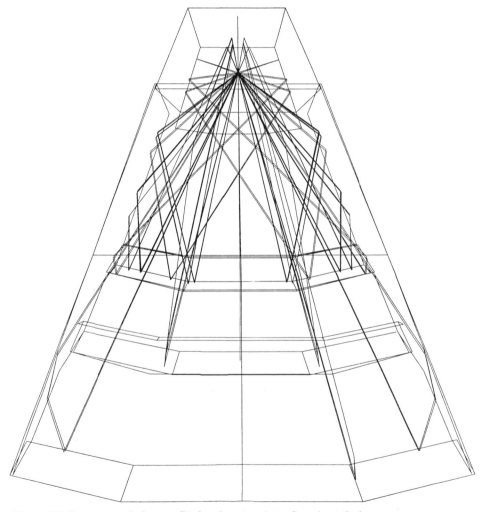

Figure 10.5 Sound ray paths in an auditorium (courtesy Artec Consultants Inc)

Intelligent buildings

Building services have up to very recently been under the control of relatively crude devices such as thermostats and timers. With the current cheapness and reliability of electronic systems, however, attempts are being made to integrate environmental control into a single system. Buildings under such control are known as 'intelligent' or 'smart' buildings[137].

Data is gathered by sensors placed about the building and passed to a computer which operates control devices as necessary to maintain optimal comfort and economy. Internal sensors may return data on temperature, lighting levels, humidity, smoke levels and so on. The external sensors may detect temperature, wind speed and direction, daylight and sunlight intensity, etc. The computer is aware of the level of activity in the building due to the time of day or the day of the week and is also aware of the thermal lag of the structure. Using this data, the simpler systems regulate the lighting and heating in rooms as necessary. More sophisticated systems also control the lifts, security systems, fire detection and control, heat transfer from one area of the building to another and even the automatic flushing of urinals.

Systems can be divided broadly into 'centralised', where a single master computer is present; 'distributed', where areas of the building have a certain amount of local intelligence; and 'autonomous', where the areas have full computing capability. The autonomous systems are especially appropriate where a number of buildings in a complex are linked, as for example in a large factory or university campus.

The architect needs to be aware of the possibilities offered by intelligent buildings so as to be able to make provision for the control systems. The cost of the multitude of sensors and more importantly of the mechanical control devices will add significantly to the cost of construction. On the other hand, the running costs of the building will be greatly reduced. A good design will use the minimum of data and control circuitry consistent with efficiency and with flexibility if the internal arrangement of the building should change. Architects must be able to handle this new aspect of design and must be able to justify its cost to the client.

The ABACUS environmental analysis system

Over the past ten years, the Architecture and Building Aids Computer Unit (ABACUS) at Strathclyde, Scotland, has been developing and refining a suite of inter-relating programs for a wide range of environmental analysis[138]. The suite consists of two principal programs, GOAL and ESP ('Environmental System Performance') which have been implemented on a number of the more powerful microcomputers running under the UNIX operating system. Over fifty systems have been sold world-wide, approximately half of which have been to educational and research establishments.

The suite is one of a number of commercially available packages, but has the advantage for the architect that it is possible to work at several levels of detail. At the sketch-design stage the architect can investigate the performance of different building shapes and at the detailed design stage make checks on a room-by-room basis. To make this possible, the data is arranged so that it can evolve naturally with the design just as an architect works from a broad concept to a detailed scheme.

GOAL and ESP are applicable at different design stages. GOAL requires minimal data and is intended to give quick but approximate results at the sketch-design stage. The building is defined as a set of rectilinear blocks and a separate standard data file gives very general information on the available constructions, the capital cost and running cost elements and the climate.

The program can output a range of results which are useful at the sketch-design stage, but the most important output from the financial point of view is the predicted annual energy consumption. *Figure 10.6* shows a typical output of this kind.

```
ENVIRONMENT  : ANNUAL ENERGY CONSUMPTION            15-aug-84 14:22 E11
GEOMETRY     : HOTEL DESIGN (GEOMETRY 1)    HOTELNEW    GOAL V 4.0
CONSTRUCTION: HOTEL CONSTRUCTION 1          HOTELN
PROJECT      : HOTEL SITE: GLASGOW          HOTELNEW
```

	WINTERSEASON		SUMMERSEASON	
	KWH	GJ	KWH	GJ
TOTAL SOLAR GAIN	−72646.	−262.26	−16038.	−57.90
CONDUCTION	234484.	846.51	20909.	75.48
VENTILATION	459553.	1659.04	40040.	144.55
OCCUPANCY	−51116.	−184.54	−22080.	−79.71
LIGHTING	−153927.	−555.69	−66491.	−240.04
TOTAL	416348.	1503.06	−43661.	−157.62
PER M2 FLOOR AREA	150.	0.54	−16.	−0.06
PER M3 VOLUME	50.	0.18	−5.	−0.02
PER OCCUPANT	4003.	14.45	−420.	−1.52

Figure 10.6 Energy consumption (courtesy ABACUS)

In order to be able to produce precise results, ESP demands far more accurate data, which is prepared as several separate sets. The first set, the Geometry file, defines the shape of the buildings and the positions of the internal walls, windows and so on. The file is itself in three sections. The first gives the broad outline by defining the buildings on or adjacent to the site as a number of blocks. The second section describes the internal details of each block giving the floor levels and the positions of the walls that enclose the rooms for which results are required. The third section defines the size and position of the windows.

The second set of data is the Thermal Properties file which defines the characteristics of the materials from which the building is constructed. The data in this file is used when calculating such factors as heat flow through the surfaces, the amount of sunlight absorbed and the thermal lag of the fabric. A lot of parameters are required, including those for density, specific heat and conductivity. The solar transmittance of the windows must also be given.

The third set of data is the Project file which gives parameters applying to the whole scheme. It includes the site latitude and also figures for incidental heat gains from such sources as the occupants and any machinery installed. The Project file typically occupies only a few lines.

There is a fourth file that contains climatological data, but this will not normally involve much preparation time as it contains a range of standard annual hourly figures for representative localities, one of which can be chosen by the user. Only if the building is in an unusual climate or if the designer wishes to test the effect of very extreme conditions will it be necessary to gather data for this file.

Not all the data need be prepared in full, and the program will make assumptions where information is missing. So at an early design stage only the first part of the Geometry file need be completed and the building regarded as a set of large block shells. From this data the overall energy consumption can be calculated for a given building form and with a given percentage of glazing. Later, more detail can be added to give a fuller and more accurate evaluation. Buildings adjacent to the site will usually be left as block shapes in the Geometry file because they do not need to be analysed but may shield some daylight or sunlight from one of the buildings being designed. The structure of the Geometry file not only allows definition at several levels of detail, but also allows buildings to be moved or re-oriented on site with the minimum of data changes. Walls, floors and windows are all defined relative to the block shells, so even at a late stage the effects of changing location or aspect can be investigated by altering a single line in the data.

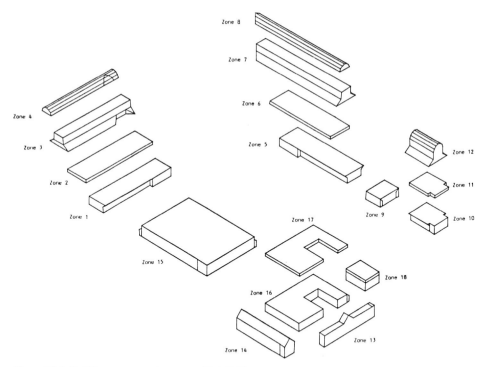

Figure 10.7 Building geometry (courtesy ABACUS)

One module of ESP is devoted to inputing and checking the data and includes the ability to output the geometry of the building. *Figure 10.7* shows such an output.

ESP is most effective in evaluating the thermal performance of the building. Dynamic analysis of heating and air conditioning is theoretically rigorous and can produce very accurate results. Daylight and sunlight analysis can be performed, but no artificial lighting results are available at present. Simple acoustic analysis is provided in the form of the calculation of reverberation times in the rooms, from which their general suitability for speech or music can be checked.

The program is normally run interactively, with the results being displayed graphically on the screen. This gives a very clear impression as the user can see quickly if a design is suitable and if necessary can modify the data files and so iterate towards an optimum solution. For precise evaluation in critical cases, results can also be presented as numerical values.

In the thermal analysis section of the package, there is a fairly large number of possible variables. The user has a number of different analysis options depending on which factors are held constant, which factors can vary between fixed limits and which factors can vary freely. At the simplest level, the user may specify that the internal air temperature be held constant. The program will use the data in the Climate file to find how the external air temperature will vary and from this can calculate the necessary plant capacity and the annual energy consumption. The internal air temperature can also be allowed to vary between specified limits and this gives an indication of the energy savings accruing from intermittent plant operation. The program can print out an operational strategy for the plant if required. The latest version of the program has various plant strategies built in. The user can test these by pressing certain function buttons on the keyboard.

The standard variations would answer most of a designer's questions, but in general any factor can be fixed or allowed to float and different combinations can be useful in investigating particular aspects of the design. For example, in extending a building, the architect might be reluctant to scrap the existing heating plant as this would be very expensive, so the plant capacity could be fixed and the results will then show how the internal temperature would vary. If the performance was not totally inadequate, the design could be changed so as to increase insulation or use smaller windows. Alterations to the Thermal Properties or Geometry file could be made to test the adequacy of this. These alterations may be made interactively.

The output from the thermal analysis can be graphical or textual. The terse textual form lists the magnitude and times of occurence of the maximum, minimum and average values calculated, while the full textual form gives the values at specified time intervals for the entire period of interest. The graphical output gives a less precise but more vivid impression of the results. At the investigative stage the variation in all the factors involved would usually be plotted superimposed on one another. At a more detailed stage, one or two factors only might be isolated, the performance of the others having been decided. *Figure 10.8* shows a typical output in which the internal air temperature is held constant during working hours, the variation in external air temperature is displayed and the necessary plant output is shown as calculated.

The user is provided with a menu of the different thermal components on the screen which can be used to choose the factors to be displayed. The way the factors behave will quickly give insight into which areas are the most significant in the overall design for the specified design.

Condensation checks are another option associated with the thermal analysis section. Checks can be performed on surfaces and at interfaces. One aspect of this is that the insulation efficiency of a construction can be studied and ways to improve its efficiency or save money become obvious. *Figure 10.9* shows the internal fabric performance of a multi-layered construction.

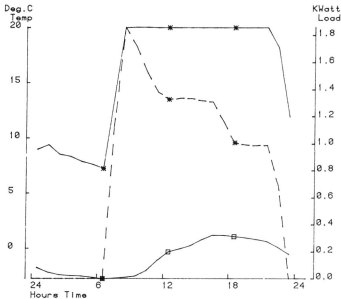

Figure 10.8 Plant capacity predicted by ESP (courtesy ABACUS)

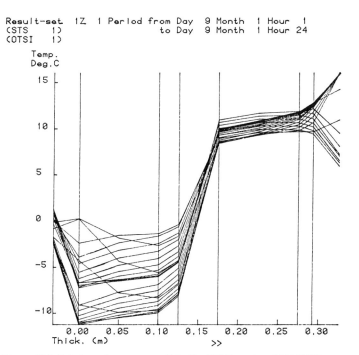

Figure 10.9 Fabric performance predicted by ESP (courtesy ABACUS)

BLOCK 3 FACE 2

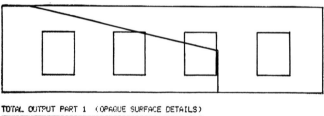

TOTAL OUTPUT PART 1 (OPAQUE SURFACE DETAILS)

LATITUDE OF SITE 51 70

BLOCK NO.   3 DAY 1 OF MONTH 1   TIME 10.00 HRS.

SOLAR AZIMUTH =   152.1
SOLAR ALTITUDE =   10.7

Figure 10.10 Sunlight reception predicted by ESP (courtesy ABACUS)

Solar gains are incorporated into the integrated results for the thermal balance and it is possible to show which parts of the building are shaded from the sun at any time and on any day. Graphically, the results are shown as a shaded elevation, an example of which is given in *Figure 10.10*.

The other aspect of natural lighting is the daylight factors within a room. To calculate daylight factors, the window and room dimensions and the data on external obstructions are taken from the Geometry file. The reflectances of the various surfaces are taken from the Thermal Properties file and the program then has enough data to make an accurate calculation. The results can be displayed as values on a grid placed over a room plan. From this the designer can immediately see if some corners of the room will be excessively dark, or if permanent supplementary artificial lighting will be required.

When energy use analysis is being performed, the program will detect at what times the daylight levels are so low that the artificial lighting must be switched on, and will incorporate this energy use into the calculations.

Chapter 11

Office management

The need for office management systems

The average architect tends to feel that his responsibility is solely to produce the best design possible within the brief, but he is also part of an organisation that has to justify the work it does financially. If the office is in private practice it is essential that the fees received exceed the cost of the work done or the firm will go bankrupt. Even offices run by national or local government bodies must take care not to use too many resources on a project as this will waste the taxpayers' money or lead to an imbalance of effort put into different jobs.

In a large firm, the responsibility for secretarial and administrative functions can be delegated to some extent to professional supervisors. However, architecture is a very fragmented profession with most practices being small; the average size of practices in the UK for example, is only five staff.In these circumstances it is inescapable that many architects must take over administrative tasks, and even architects employed in large firms should have at least some familiarity with the problems so as to be able to organise work efficiently.

Using traditional methods, these tasks are very onerous and demanding. Fortunately, most secretarial and accounting problems are ideally suited to the computer as there is little or no need for arbitrary decisions and the methods of obtaining the results and the way they must be presented have been standardised for many years. A microcomputer and a few programs dealing with these functions is one of the most cost-effective investments a small office can make.

Job-cost analysis

One of the most important administrative problems in an architect's office is to keep track of how much time is being devoted to each job so as to be able to identify those that are running over budget.Architectural practices and most other professional organisations get most of their income from a large number of separate contracts with a relatively short life.This is in contrast to the more usual business situation where a manufacturer, say, has a certain number of outlets which do not often change.

Most offices require their staff to fill in weekly 'time sheets' which record the number of hours spent on each job. The administrator then multiplies these hours by a cost which will vary according to the salary of the members of staff involved and totals the figures to give the cost on each job.

Programs to automate this process are easy to produce and a wide range is available for microcomputers. As their task is rigorously structured, they all operate in a broadly similar way. A master list of employee names and their hourly rates will be stored on a disc file and there will also be a master list of job names and numbers. When the program is run, usually weekly, the system will typically display each employee name on screen and allow the administrator to type in the reference numbers of the jobs worked and the number of hours spent on each job from that person's timesheet. The validity of the job number will be checked and the appropriate cost set against the job. Provision will be made to distinguish between normal working and overtime and for holiday and sickness leave[139].

Most programs will also provide means to enter costs against jobs from submitted expense sheets and may provide means to enter interim fees received to be put as credits.

Various outputs will then be possible. The most useful of these will be a cost summary broken down by job. *Figure 11.1* shows an output of this kind which was produced by the JC-MICRO program distributed by ABS Oldacres Computers Ltd. and which is available on a wide range of microcomputers.

This output gives a short overall summary for each job for management purposes. Total hours, costs and expenses for the jobs are shown in the first group of columns. A second group compares progress to date against total cost and expected income and the third group of columns shows the last fee amount invoiced, the total fee and the last fee date.

The JC-MICRO program provides various other outputs, including more detailed outputs for each job specifying the employees who have worked on the job and the dates between which they were working, and a condensed management summary consisting of a single line for each job for general checking of progress.

Accounting

Most architectural firms, being partnerships rather than companies with shareholders, are not legally required to publish formal accounts. It is, however, obviously necessary to have a clear statement of the financial position of the firm so as to be able to monitor performance and also for tax reasons.

The largest expenditure in a professional firm, whose assets are mainly the skills of its staff, is the payroll. The task of calculating pay from hours worked, salary, tax, amount of pension contribution due and so on is a long but mechanical process which can easily be computerised. Basic information on all members of staff can be kept in one file on disc which will only need to be updated occasionally. Calculation of the pay can then be fully automatic. Many computer programs are available to carry out this task.

The most active parts of an accounting system are the Sales Ledger and the Purchase Ledger. The Purchase Ledger records what items and services were bought to keep the firm running. A manufacturing concern would normally be recording large sums for raw materials in this, but the typical architectural practice will be spending relatively small amounts for equipment, drawing supplies and so on.

The Sales Ledger records the income. In a professional practice this will consist almost entirely of fee income and so is usually called the Fee Ledger.

MANAGEMENT SUMMARY

SEPTEMBER 1986   5/9/86   PAGE 1

JOB NR	JOB NAME	THIS MONTH	CUMULATIVE	BUDGET COMPARISON	FEE CLAIMED	DATE LAST COST ENTRY
JEC741	Javaid Engineering Company					
	ALLOCATED F : 16 H : 18 N : 40 R : 20 PTNR : JCH ASSOC : CJH FEE 96900 (S			250000	38.76%) BUDGET 65%	
	COMPLETE : 100 80 90 85 T : 85 6 48					
	STAFF COST :	260	52190	PROGRESS : 54343	THIS MONTH : 344	
	EXPENSES :	0	1740	TO COMPLETE : 8642	DATE LAST ALL	
	STAFF HOURS:	18	4442	GROSS EARNING: 83605	0	10/09/86
	NR EXPENSES:	0	1809		38154	
KHD106	K.H.Davis Ltd.					
	ALLOCATED A : 15 D : 12 M : 35 N : 24 Q : 99 PTNR : JCH ASSOC : FT FEE 86500 (S			86500	100.00%) BUDGET 76%	
	COMPLETE : 100 100 89 99 14 97					
	STAFF COST :	0	56400	PROGRESS : 62775	THIS MONTH : 4948	
	EXPENSES :	0	2886	TO COMPLETE : 2965	DATE LAST ALL	
	STAFF HOURS:	0	4347	GROSS EARNING: 82599	0	06/08/86
	NR EXPENSES:	0	1427		43045	
PEM001	Pascals Estate Management Ltd					
	ALLOCATED A : 30 B : 25 H : 15 N : 15 R : 15 PTNR : JCH ASSOC : SSE FEE 82000 (S			0	0.00%) BUDGET 80%	
	COMPLETE : 100 100 100 80 15 95					
	STAFF COST :	0	49480	PROGRESS : 63140	THIS MONTH : 11615	
	EXPENSES :	0	4910	TO COMPLETE : 2460	DATE LAST ALL	
	STAFF HOURS:	0	4402	GROSS EARNING: 78925	0	06/08/86
	NR EXPENSES:	0	2045		30538	
SCF091	Southwick – Clearwater – Footman					
	ALLOCATED A : 25 B : 25 D : 25 N : 10 S : 10 PTNR : JCH ASSOC : CJH FEE 135000 (S			300000	45.00%) BUDGET 70%	
	COMPLETE : 100 70 40 10 20 18					
	STAFF COST :	220	30000	PROGRESS : 49234	THIS MONTH : 17730	
	EXPENSES :	0	4090	TO COMPLETE : 45266	DATE LAST ALL	
	STAFF HOURS:	17	3057	GROSS EARNING: 70335	0	10/09/86
	NR EXPENSES:	0	1504		19994	
SSS934	Super-Safe Security Services					
	ALLOCATED B : 15 C : 12 D : 24 N : 40 S : 10 PTNR : JCH ASSOC : SSE FEE 45000 (S			45000	100.00%) BUDGET 76%	
	COMPLETE : 100 90 65 10 9 0					
	STAFF COST :	0	13730	PROGRESS : 15527	THIS MONTH : 1030	
	EXPENSES :	0	3070	TO COMPLETE : 16673	DATE LAST ALL	
	STAFF HOURS:	0	1019	GROSS EARNING: 20430	18/09/86	06/08/86
	NR EXPENSES:	0	767		10000	
					11405	

Figure 11.1 A job summary report (courtesy ABS Oldacres Computers Ltd)

Computer programs are available to handle Sales Ledger and Purchase Ledger tasks separately or as an integrated package[140, 141]. The user is prompted for the name of the supplier or client and the amount involved and the computer will make an entry on a disc file together with the current date. It is also necessary to associate a code with the transaction. Thus postage costs might have one code, heating and lighting costs another and so on.

At intervals, typically monthly and yearly, it is necessary to total the sales and purchases under various headings. In normal practice, this is done by keeping a 'Nominal Ledger' in parallel with the Sales and Purchase Ledgers. The Nominal Ledger is divided up by categories or by the names of specific large firms, and entries from the Sales and Purchase Ledgers are copied into these categories. The use of a computer program can eliminate the keeping of a separate Nominal Ledger as the transaction codes allow automatic categorisation of entries.

The Nominal Ledger summary forms the main part of a standard presentation of the financial situation: the Profit and Loss Account. To the amounts spent recorded by the Purchase Ledger must be added the staff salaries and the depreciation of fixed assets such as office furniture. The fees invoiced as recorded by the Sales Ledger form the income, and the excess of income over expenditure over a fixed period is noted.

The other important presentation of the financial situation is the Balance Sheet. This is essentially a statement of how much the firm is worth. On the credit side is recorded the value of the fixed assets such as furniture and equipment and the value of the freehold or lease of the office itself if the practice owns it. The credit side also includes the cash and investments the office has and any fees owing. On the debit side are the current liabilities, usually unpaid invoices. The difference between the sums gives the net current assets of the practice.

A large integrated accounting package can produce Profit and Loss Accounts and Balance Sheets automatically. However, this is often not worthwhile as it requires the input and checking of data which is not actually processed, but merely printed out again on a statement. Most efficient working may be obtained by the use of separate payroll and Sales and Purchase Ledger programs, then using the results to prepare statements by hand.

Word processing

Although drawings are the main form of communication, an architect's office still generates a huge amount of typewritten information, including letters, Architect's Instructions, specifications, memoranda and so on. In a large office the typists will typically make up ten to fifteen percent of the staff. Techniques generically known as 'word processing' have been developed to aid typing and controlled tests have shown that productivity can be increased by two or three times the manual equivalent. Word processing is therefore a very cost-effective application in any office[142, 143].

In essence, word processing involves sending the typed information to a disc file rather than directly to paper. This gives several advantages. Most importantly, corrections can be made with little effort. With manual methods, slow and messy removal of text with erasing fluid is necessary, then the correct version must be overtyped. In the worst case, if a line has been omitted for instance, the entire page may have to be retyped. Word processing allows corrections to be made on the computer screen and the remaining text will adjust itself to accommodate insertions or deletions. Corrections are therefore very fast.

An extension of this facility means that standard letters can be kept on disc files and printed out at any time with the necessary changes.Thus standard letters to tenderers, local authorities or applicants for employment can be devised and output when required with corrected names, addresses and project identifications. Another important facility enabled by word processing is that blocks of text can be moved around in a document or copied from one document to another. In an architect's office this is especially useful in the preparation of specifications. A standard complete specification can be held on one disc and paragraphs extracted and corrected as required to build up a specification for a particular job. This is far quicker and more accurate than the traditional manual techniques of cut-and-paste methods followed by complete retyping.

Another use of the facility is keeping a file containing the names and addresses of clients, contractors and other organisations written to regularly. The appropriate block of text can then be copied onto the head of a letter and onto a gummed address label as required. This alone can save hours of typing every month.

Correction, moving and copying text form the main strengths of word processing systems. They do however, offer other facilities which are often useful. One such facility is the ability to search for a name or other sequence of characters. Thus the user can search for a paragraph number in a long document or for the word 'architect' in a specification, for example.

Searching can be combined with automatic replacement of text. So if the supplier of some item were to be changed on a specification, the user could instruct the machine to replace all occurrences of 'ABC Corporation' by 'XYZ Corporation', for example.

Some machines can sort lines into alphabetical or numerical order. Thus if a list of clients is kept it can be sorted into alphabetical order for easy reference. If sorting is available, a word processing program can be used as a simple replacement for a tabular database. For example, if a list of room numbers and equipment contained is typed in, it is possible to copy lines where rooms are the same, to alter them where rooms are similar but not identical, and to sort on any column to obtain schedules for different purposes. It may well be easier to do this than to train someone in the special techniques of a database management system.

Another important facility that is often provided is a spelling checker.The computer will contain a list of typically 30,000 to 100,000 words. Every word in a document can be checked against the list and if it does not appear, that word will be marked as a possible mis-spelling. Many marked words will be proper names and the computer is unable to help if the typist uses an incorrect word such as 'principle' when 'principal' is meant, but this facility will trap most errors.

Some systems allow the user to add a list of words specific to his own applications. An architectural practice might include a range of technical terms which would not appear on a conventional list.

An incidental, but important, advantage of word processing is that the physical results are always perfect. Corrections will not be visible; there will be no misalignments and if required, the text can be right-justified, that is, with an even right-hand margin, as in this book.

The user of word processing is faced with a choice of implementation. By far the cheapest method is to buy a word processing program and load it onto a standard microcomputer. Dozens of programs are available for very small sums. An alternative is to buy a dedicated word processing machine. This is more expensive, but because it is geared to a particular application is faster and easier to use. Most importantly, special keys will perform particular functions and will be labelled as such. A general-purpose machine might require a typed command or use unlabelled function keys.

Some dedicated machines are able to display on screen exactly what will be output on the printer. The screen itself is turned through ninety degrees so that the proportions approximate to the printed page, whereas a general-purpose machine will always have the screen in the conventional orientation with the long edge horizontal. In the latter case, it is normally only possible to show twenty-four lines of text at once, so two or three screenfuls will make up a printed page.

The best systems allow underlining of text, italic text and 'boldface', that is, heavy typing, to be shown on screen. It is obviously much easier to use a system where a page can be checked on screen before it is printed.

If a dedicated machine is decided upon, it is then necessary to decide if it is to be single-user or multi-user. The multi-user solution is cheaper as the typists can share a high-quality printer, a fast and high-capacity disc and a fast central processing unit. It has the disadvantage that any fault will put all the typists out of action instead of just one. *Figure 11.2* shows a popular multi-user system.

Figure 11.2 A multi-user wordprocessing system

Of the many programs available for general-purpose microcomputers, by far the most popular is 'WordStar'[144] which is supplied by MicroPro Inc. The latest version, WordStar 2000, currently has about 24% of the world market share. It has many advanced features, including a set of menus to guide the user through the functions available and the ability to store frequently-used phrases and add them to the document by typing a short code.

WordStar has an associated program called 'CorrectStar' which is its spelling checker. This contains 65,000 words and can be added to by the user as required. A useful facility of CorrectStar, especially for poor spellers, is that if a word is not recognised, the computer will suggest a list of words which the user might have meant. This is done both on the basis of changing one character for another and also on the basis of words that would sound rather like the word typed.

Business graphics

As every architect knows, drawings are a far better way of conveying information than written descriptions or lists of figures. This rule applies in management techniques as well; a clearly constructed graph or diagram will convey the facts more quickly and more vividly.

The problem has been in the past that to work out and then draw a diagram neat enough for formal submission is likely to take an hour or more and it is often not possible to spare this time. Fortunately, microcomputer programs are now widely available that can take a short list of figures and automatically construct a diagram from them. The diagram will be displayed on screen and can be output on a hard copy unit. The results are often a little crude because of the limits of screen resolution, but are neat and clear[145].

The three most popular forms of presentation are the graph, the histogram and the pie diagram. Each of these will be used in different situations. The graph is best at showing a single factor varying over time. For example, *Figure 11.3* shows a computer-produced graph of office overheads over a year. It is easy to see the large increase in costs in the winter months due to heating bills.

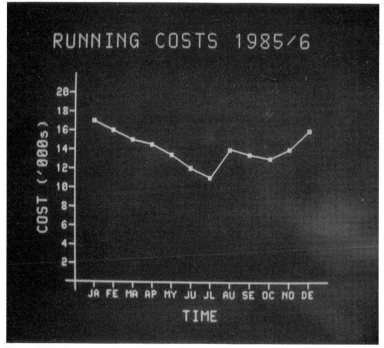

Figure 11.3 A computer-produced graph

The histogram has very similar uses to a graph but is rather clearer when more than one factor has to be displayed. For example, *Figure 11.4* shows the fees and expenditure of a practice for the four quarters of one year. It is immediately clear from this that the number of commissions has not varied much, but that the office has increased its efficiency over the year.

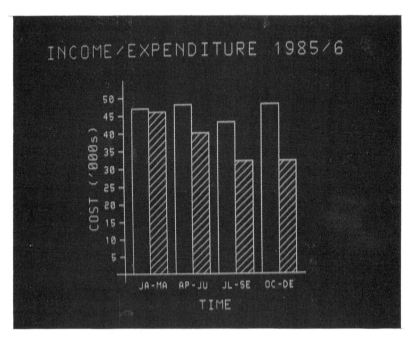

Figure 11.4 A computer-produced histogram

The pie diagram is used to show how a single quantity, usually of money, has been divided up. *Figure 11.5* shows how the office expenses are proportioned. This gives a very clear impression of which expenses are the heaviest and therefore where to concentrate cost-cutting measures.

Using a computer program, the graph required the typing of just 12 numbers plus labels for the axes and the title. The histogram needed eight numbers and the pie diagram six. The results were obtained in a few minutes, but manually the set would have needed one or two hours by the time the scales were worked out and toning and stencilling added. In practice, they would probably not have been produced at all manually and any decisions taken would be on the basis of a list of figures.

Spreadsheet programs

Spreadsheet programs are a comparatively recent development. The first spreadsheet, 'VisiCalc', was launched in 1979. It became the best-selling applications package ever, and still remains popular in its later versions.

A spreadsheet can be visualised as a large gridded sheet of paper. Each square of the grid, called a 'cell', can contain text, a number or an arithmetic expression. The arithmetic expression can use numbers to be found in other cells and it is from this ability that spreadsheet programs derive their power[146].

Conventionally, the rows of the grid are numbered and the columns lettered, thus the cell A1 is the top left-hand cell and so on. A simple example showing the way the system works can be given by setting cell A1 to the number '10'; setting cell B1, which is immediately to the right, to the number '6' and setting cell C1 to the expression 'A1-B1'. When the spreadsheet is in operation, cell C1 will display the number '4' which is the result of the calculation. However, the expression itself remains in force and if either A1 or B1 is changed then C1 will change automatically.

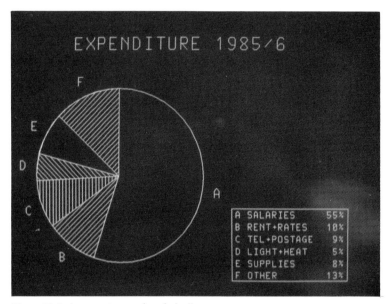

Figure 11.5 A computer-produced pie diagram

In practice, many figures and expressions will cover the spreadsheet. The user can then roam across the sheet and try the effect of various alterations. He might for instance alter a figure giving the exchange rate between currencies. As the change is made, many figures in other cells may be seen to alter because they are connected with the exchange rate. Thus different scenarios can be tested, or the effect of making changes can be evaluated. The spreadsheet has become one of the standard modern management tools and millions of copies of such programs have been sold.

A simple example drawn from architectural practice can be given. A question that has to be asked regularly is: 'Should the office employ more staff, or take longer to complete jobs?' The following spreadsheet might be made up to investigate the answer.

	A	B	C	D
1	Job Name	Cost	Fee	Staff Time
2	11 Long Lane	100,000	B2*B10	3.0
3	21 Park Avenue	150,000	B3*B10	4.5
4	53 London Road	60,000	B4*B10	2.5
5	81 Acacia Cres.	500,000	B5*B10	10.5
6	2 Oak Street	75,000	B6*B10	2.0
7				
8	Totals	sum(B2:B6)	sum(C2:C6)	sum(D2:D6)
9				
10	Fee percentage	0.06	Overheads	5000
11	Staff numbers	5	Staff salary	1000
12	Expenses	D10+B11*D11	Elapsed time	D8/B11
13	Gross profit	C8−D13*B12		

This might show the situation in a small office. A list of job names is entered into column A, their construction costs into column B and the amount of staff time each will take in man-months into column D. Column C is made up of expressions which take the construction costs in column B and multiply them by the fee percentage which has been entered in cell B10. Thus the fee will be calculated automatically for each job.

Row 8 holds the totals. The entries here show the use of a 'function' which is a shorthand way of writing certain commonly-encountered expressions. In this case the function 'sum' totals a row or column of figures, thus the expression 'sum(B2:B6)' is equivalent to 'B2+B3+B4+B5+B6'.

Other cells hold various constants. The current number of staff; the monthly office fixed overheads such as rent and heating, and the average monthly staff salary are held in cells B11, D10 and D11 respectively.

Other cells perform further calculations. Cell B12 calculates the cost of running the office each month by multiplying the number of staff by the average salary and adding the fixed overheads.

One cell summarising the position is D13 which calculates the elapsed time to carry out the work by dividing the total number of man-months in D8 by staff numbers in B11. Cell D8, of course, is the result of a calculation in itself.

The other summary cell is B13 which finds the gross profit by taking the total fee income in C8 and subtracting the monthly expenses in B12 multiplied by the number of necessary months in D13. In this case, several steps of calculation have been necessary to arrive at the final figure.

In action, the expressions are invisible and only their results are seen. Text and fixed numbers are always visible. Text may be placed freely in cells as titles and comments, but of course calculations will not normally involve such cells. The example when running will look as follows.

	A	B	C	D
1	Job Name	Cost	Fee	Staff Time
2	11 Long Avenue	100,000	6000	3.0
3	21 Park Avenue	150,000	9000	4.5
4	53 London Road	60,000	3600	2.5
5	81 Acacia Cres.	500,000	30,000	10.5
6	2 Oak Street	75,000	4500	2.0
7				
8	Totals	885,000	53,100	22.5
9				
10	Fee percentage	0.06	Overheads	5000
11	Staff numbers	5	Staff salary	1000
12	Expenses	10,000		
13	Gross profit	8,100	Elapsed time	4.5

The manager can now see how long it will take to carry out the work (4½ months) and what the gross profit will be. He might then try altering the number of staff in B11. This will instantly change the values given for monthly expenses, the elapsed time and the gross profit. The manager can test other possibilities by altering the values; perhaps adding a likely job to the list and so on.

In practice the spreadsheet will be much more sophisticated than the example, especially in allowing jobs to start at different times. A large office will have dozens of jobs running simultaneously and in such circumstances spreadsheet methods will save days of analysis and large sums of money.

All spreadsheet programs impose a limit on the number of rows and columns that may be present. For practical use at least 500 rows by 100 columns should be available, but many programs allow thousands of rows and columns.

Spreadsheet programs also tend to be greedy for main memory. It is important to check if extra memory must be purchased to run the program.

An important feature of practical spreadsheet programs is that expressions can be generated without repetitive typing. For instance, the repeated expressions in column C in the example should not have to be typed individually.

Another important feature is the ability to move blocks of cells around the grid and have the expressions alter automatically. For example, if it were decided to move the staff numbers in cell B11 elsewhere, then all expressions using B11 should change automatically to reference the new cell.

Integrated business systems

In the last few years, a number of firms has developed integrated packages that combine several business application programs into a single system.

Two of the most well-known of these are 'SYMPHONY' which is supplied by the Lotus Development Corporation and 'FRAMEWORK' from Ashton-Tate Inc. Typical applications that are handled are word processing, database management, spreadsheet analysis, business graphics and appointments diaries[147].

When entering such a system, the user is typically confronted with a 'master menu' that offers the choice of working in one of the application areas. In that area, say word processing, another menu will appear that offers a choice of functions to carry out that application. At any time, the user can exit from the application area back to the master menu and from there either carry out another application or exit from the program entirely.

The drawback of integrated business systems is that the user must accept the system as a whole with its strengths and weaknesses. Thus while a given system might have excellent spreadsheet, say, it might only support a small and restrictive database management program. If all functions must be carried out in the best possible way it is preferable to buy a range of programs each of which is the best in its particular field.

The great advantage of integrated systems is that information can be moved very easily from one application area to another and this is often a very powerful facility. To take an example, the user might store information on jobs in the database. This can be interrogated and updated in the usual manner by the database management section. It might then be useful to perform spreadsheet analysis using certain fields in the database. When using separate programs, the transfer of data would be a difficult task in itself, probably needing several hours to reorganise the information output by the database program into a form acceptable to the spreadsheet program.

When using an integrated system, it is possible to transfer the information in a field in the database directly into a specified column on the spreadsheet. Having performed the analysis, the user might wish to incorporate the figures generated into a report. This is easily done by transferring a block of cells from the spreadsheet into the word processing section. The user can then add explanatory text around the essential figures and produce the report. Similarly, figures from the database or the spreadsheet might be transferred into the business graphics section and graphs or histograms produced to illustrate the report.

The other advantage of integrated systems over separate programs is that the commands needed to work each application area will have a resemblance and so using the system will make fewer demands on the user. As a simple example, some programs use the 'backspace' key to cancel the last character typed while others use the 'rubout' key. In an integrated system the conventions will be the same throughout. At a higher level, the keys used to activate and select from a screen menu will be the same. A mental effort has to be made when switching between separate programs so as to adapt to a rather different way of working.

Many integrated systems allow the user to display several results on the screen at once. This method of presentation, known as 'windowing', means that the user can keep track of more than one aspect of the data at once. Thus a list of numbers and its corresponding graph can be seen together. This is often a convenient way of working, although it has to be said that because of the limits of size and resolution, a screen will become cluttered and unreadable fairly quickly.

Chapter 12
The future

It is dangerous to prophesy on technological matters as they seem to have a disconcerting knack of getting their own back. History is littered with predictions that, with hindsight, seem ridiculously short-sighted, and these have come from even the most eminent men. For example, Professor Simon Newcombe, the director of the US Naval Observatory, wrote in 1884 that, 'Artificial flight is impossible'[148]. Twelve years later, the Wright brothers were proving him wrong. Lord Rutherford, probably the most important of the early workers in nuclear physics, said in 1933 that, 'Anyone who expects a source of power from the transformations of these atoms is talking moonshine'[149]. Twelve years later, he was proved wrong in the most emphatic fashion. In 1956, Dr Richard Wooley, the UK's Astronomer Royal, dismissed the ideal of space travel as 'utter bilge'[150]. The following year, Sputnik I went into orbit.

The track record is no better in the more specific field of predicting developments in computing. In 1948 the expert opinion was that one machine would suffice for all Britain's computational needs: today there are tens of thousands of installations. In the 1950s an expert pointed out that computers could never handle complex problems because the thermionic valves they would need for high performance would keep burning out at an impossible rate[151]. Even as he was writing, the first transistorised computers were being marketed.

In more recent times, the blindness to the way things were going seems incredible in retrospect. I cannot find a single prediction in our own company's clippings files prior to 1970 that minicomputers would ever be important; the expert view was that costs would drop by building larger and faster machines and that users would communicate with these by means of individual terminals using the telephone network. Just a few years later, minicomputers became, and remain, one of the most important sections of the market. The accepted view then became that eventually every office and household would have its own machine. Since then, the development of microcomputers has caused yet another revolution in expert opinion and it is now thought that each individual will have control of a range of computing facilities.

If the most eminent scientists and thinkers can get things so badly wrong, there is obviously little point in placing any reliance on predictions, including those incorporated in this book. It is better to make decisions on the basis of the situation as it exists at any given time and not to try to anticipate developments in technology.

About the only statement that can be made with confidence is that wherever things are going, they are going there fast. Hundreds of new products in the computing field are launched each month, many of them highly sophisticated. The problem at present is that it is impossible to keep up with this rate in harnessing the power of these devices; we now have much more machinery than we know how to control.

Computing is probably the fastest-developing technology that has ever existed. The first fully electronic computer, named ENIAC, was built by the University of Pennsylvania in 1945. It contained 18,000 valves, occupied 90 cubic metres of space and used 200 kilowatts of electricity. It is now possible to buy for a very small sum a computer that can be held in the hand and consumes less power than a light bulb, but which is twenty times faster than ENIAC and contains the equivalent of over 500,000 separate circuit elements. Some idea of the miniaturisation of complex circuits that is now possible is given by the picture of an integrated circuit in *Figure 12.1*. The circuit is about 2 mm square and contains the equivalent of 2,600 transistors. The photograph was taken through a Philips scanning electron microscope.

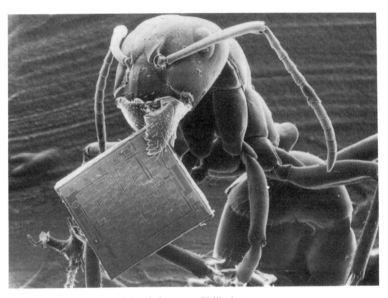

Figure 12.1 An integrated circuit (courtesy Philips)

There is no reason to suppose that any sort of limit has been reached, and even more impressive comparisons will be possible in a few year's time, but the devices that now exist are already cheap enough and powerful enough to replace many crude electromagnetic controllers. This process is already well advanced, and many familiar machines that were previously controlled by a tangle of wiring, relays and clockwork can now be supervised by a small electronic package, with considerable gains in cost, reliability and effectiveness. These machines include lifts, which can have very sophisticated strategies to maximise efficiency under varying conditions; traffic lights, which can maintain optimum flow throughout the day, and household appliances such as washing machines, cookers and sewing machines which can offer a wide range of automatic programmes.

We have also seen the emergence of completely new machines which have not previously been economic to manufacture, but which with microprocessor control can be made cheap enough to capture a large market. These include personal computing devices like pocket calculators and microcomputers, digital watches, language-translation machines and entertainment devices that can play games such as chess against a human opponent. On a commercial scale, robot arms that carry out tasks such as welding and paint-spraying on an assembly line are already widespread.

Many other machines will be made practicable by the use of microprocessors. Some which are at an advanced stage of development are electronic diaries which will keep a record of appointments and facts and will be able to retrieve them as required, translating telephones that will take in a sentence in one language and emit it in another, and security devices that will recognise authorised people by their faces. There are many such applications of cheap computing that spring to mind and many more that will be 'obvious' only with hindsight.

Microprocessors have also had an influence on the design of full-sized computers. Many of these now hand over certain functions that were controlled by the central processing unit to microprocessors, each of which can deal with a separate part of the total problem. Thus if the computer were playing chess, each microprocessor might be investigating a possible move. This concept of working in parallel, rather than taking one step of calculation at a time as computers have traditionally done, is a most important one. Parallel processing has been implemented before on a limited scale, but has always been very expensive. With cheap circuitry available it will probably become the preferred method of working owing to the greatly increased power of the machine. One result of this, however, will be that programming languages will undergo a radical upheaval to get the best out of the new machines. One programming language designed with parallel processing in mind, called 'OCCAM'[152], has already made its appearance, and the next decade may see the demise of traditional languages such as COBOL and FORTRAN.

In summary, it seems that in the next few years there will be a great proliferation of 'intelligent' machines that will be able to do a great deal more without human control than their present-day equivalents, if any. This increase in automation will reach into every part of our lives and will eventually require a radical reorganisation of working methods in almost every field. This reorganisation will be necessary because some of the tasks to be performed will be different and because the methods of carrying out tasks will also change. Architects may be affected more than the other members of the building team because they are less familiar with the concepts involved and have lagged behind in technological development. It is not quite fair to say that the profession is still using methods that would be perfectly familiar to Sir Christopher Wren. There have been many improvements, especially in the field of communications with the introduction of dyeline machines and photocopiers, but architecture is still very labour-intensive.

The problems facing the architect can be identified and it is fairly safe to prophesy that there will be technological solutions to the worst of them. The precise nature of the solutions cannot be predicted because of the many possibilities that exist, some of which cannot even be guessed at, but, on the understanding that it is for entertainment rather than for information, some aspects of a possible scenario can be suggested.

Perhaps the most pressing problem in any office is the amount of drawing that has to be done. In the last analysis, every line in the design drawings must be individually drawn by hand, and must often be drawn several times over if changes are made, or if plans are required at different scales. The inherent inefficiency of this necessitates a large draughting team with the attendant difficulties of communication.

The increased use of computer-aided draughting will greatly ease this problem. A practical microcomputer draughting system currently costs about a years' salary and can almost double an architect's drawing speed. As they become cheaper and faster, the economics of the situation will become even more compelling. On this basis, a typical office in ten years' time will have microcomputers where drawing boards presently stand.

Once such machines have been installed, it is natural and obvious to extend the programs to analyse the drawings and produce various results. Many draughting systems can already associate lines on the drawings with textual information and so printed schedules can be produced of doors, windows, room finishes and fittings. This facility will ease some of the most tedious tasks an architect has to undertake.

Two-dimensional drawings are usually sufficient for schedules, but most other analyses require a three-dimensional model of the building to be defined in the computer. A relatively simple model will enable automatic visualisation to be performed. Once perspective views are available without effort, architects may come to use them as an important design aid. This is not the case at present, largely because only expensive machines are capable of producing results in a reasonable time. Designers may in the future spend time 'walking around' and 'walking through' their buildings to check that they have the appearance and the capability required. Colour screens and printers are already reasonably cheap, and we can expect that the views produced will be coloured and shadowed representations rather than the line drawings which are most common at present.

If more detail is given on the materials used, other results are obtainable. Perhaps most usefully, a detailed bill of materials can be printed. This will greatly speed the costing of the building. It may be that there will be a drastic reduction in the numbers of staff employed by cost engineers and quantity surveyors and a shift in their role towards advising architects on what methods and materials should be used in construction.

Given the materials of which the building is composed, environmental analysis, including heat loss, daylight factors and sunlight penetration, can be produced almost automatically. Duct sizing and the routing of services can be optimised by existing programs and will reduce the part played by the mechanical and electrical services engineer. Generally speaking, the use of computers will bring more of the design process under the architect's control and the continual erosion of his role by specialist contractor and consultants will be halted or even reversed.

Somewhat surprisingly, there is no sign at present that the use of computers will allow the architect to take over some of the tasks performed at present by the structural engineer. It might be thought that as all the data on column and wall dimensions and span lengths can be obtained automatically from a draughting system this would be merely a matter of evaluating the appropriate formulae, but in fact I know of no architectural practice that is attempting to take structural calculations beyond broad indications of feasibility. This may be because structural engineering is not such a rigid and formalised process as the architect tends to believe, but requires many value judgements and the trading-off of some factors against others.

The use of interactive draughting systems will give the greatest increase in productivity, but there will still be a lot of scope for textual and numeric processing. Most of this will aid the administration of the office by the use of database retrieval of information; word processing of letters and specifications and accounting aids. Each office will build up its own database of information including product data, research papers, technical articles etc., in the same way that they support a conventional library at present. The availability of independent databases will also increase. The architect will be able to access these for a small charge through the telephone system.

If more than one member of the building team uses computers, there are obvious advantages in being able to transfer information automatically between them. It is already common for land surveyors to collect their information on computers and to transmit their drawings to an architect's machine. This tendency is bound to increase, and there will probably be regular transmission of drawings between the members of the building team. An agreed international standard for the transmission of drawings is needed before this can become fully operational, but work is well advanced on such a standard[76].

One difficulty with the transmission of drawings is that the amount of data they contain is so high that it is not really practicable to send them over ordinary telephone lines. They can of course be sent as magnetic tapes through the post or by the rental of special fixed lines from the telephone authorities which can operate on very high rates. The UK, for instance, has a network of high-speed telephone links called the 'X-Stream' service[153], which can operate at up to 2 million bits per second.

The concept of the 'distributed office' has been current for some years. In this system, each architect would work from home on a microcomputer and communicate with a small central office and his colleagues by telephone. This has obvious advantages: the central office could be small and well-located for use in coordination and client meetings, staff would not waste time and money in commuting and female architects might also like to work from home when they have young children to care for, instead of their valuable training being lost to the profession for years, as at present.

The inadequacy of standard telephone links is the reason why this will not be practicable for some time. The telephone companies are, however, currently replacing existing copper cables by optical fibres which have a far higher capacity and it may be that high speed dial-up data transfer will be commonly available in ten years or so.

As an experiment, the French Post Office has installed optical fibre links to 1,500 households in Biarritz. Each home has a video camera and a television that can be linked into the telephone and users are able to see each other as they talk. The current intention is to have 3.1 million homes connected by the end of 1988[154].

Looking at the computing scene more generally, there is a number of problem areas where development can be expected. The problems of cost, speed and reliability are the areas where progress has been most dramatic in recent years and much further improvement can be expected. However, reasonably fast micro-computers are already well within the reach of any practice. With the use of solid-state circuitry, the processing parts of the system are already very reliable: it is the peripheral devices that, being largely mechanical, are liable to break down.

Magnetic discs are the worst offenders. Spinning at high rates and having very close tolerances, they are expensive to build and are not particularly reliable, and yet they are a vital part of any practical computer installation. Because of these problems there is considerable interest in alternative forms of mass storage to replace magnetic discs. At one time it was hoped that a development called 'magnetic bubble storage' would provide this. The principle of magnetic bubble techniques is that in certain materials it is possible to insert regions of reverse magnetisation which can be moved about by electric charges. The presence or absence of regions can therefore be used to hold binary data. Being solid-state, magnetic bubble storage is very reliable but unfortunately has proved to be too expensive given the constant increase in the amount of data that users wish to hold. These devices are used mainly in specialised areas where ruggedness is essential, such as military applications.

Other solid-state mass storage devices are under development but are some way from commercial manufacture. It may be in the end that mass storage is provided by a piece of random access memory similar to that used by the computer as its main memory, but removable. One firm has announced plans to produce a single integrated circuit capable of storing seven megabytes of information, and this is seven times as much as a standard floppy disc will hold[155].

A mass-storage technology which is rapidly becoming popular is the optical disc[156]. Optical discs are commonly 12 inches in diameter and are spun at high speed as with magnetic discs. However, reading and writing information is done by lasers. The earliest optical discs were not erasable: the laser physically burned pits or raised blisters on the surface to write information. The more recent versions use the heat of the laser to alter reflectivity of points on the disc surface and so are erasable. Optical discs are more reliable than magnetic discs because they do not need a read/write head flying very close to the surface. However, their main advantage is their huge capacity. Optical disc storage is measured in gigabytes, one gigabyte being 1,000 megabytes, and current systems offer from one to two gigabytes per disc. Optical disc systems are relatively expensive, but are becoming popular as large storage capacities are essential in certain applications, notably the storage of photographs.

Printers and plotters are other widely-used peripheral devices that are largely mechanical in operation. Reliability is not such a problem as they are slow and imprecise compared to discs. They are noisy, however, and take time to produce results. Electrostatic printers and plotters that are very fast and have few moving parts are already widely used and as reliance on the computer increases there is bound to be a move to such machines despite their higher cost.

The main output device is still the screen, and there is a strong demand for larger screens. Many manufacturers are working on the development of cheap liquid crystal display (LCD) screens and in a few years we can expect to see screens of up to A2 size, double the present maximum, which will be however only a few centimetres thick.

On the input side of the system, there is a search for mechanisms that can collect data from the user more quickly and efficiently. Almost all input at present is by typing, or by pointing at a position on the screen. Typing is of course very slow and pointing is limited in its use. Voice input is likely to become a practical method of defining data in the near future. One firm has produced a voice-operated device that is a direct replacement for the keyboard of an IBM PC/AT. The device will recognise up to 160 different words and transmit the equivalent string of characters to the computer[157]. Systems such as this must face the problems of different users' accents and of different words sounding alike, but these problems are now well understood.

Defining graphical input, as architects must constantly do, is currently done by a combination of typed commands and data and screen pointing. There is little sign at present of any radically faster approach that could be used. Three-dimensional digitisers are available and can obviously speed up the definition of models, and experimental systems exist that allow the user to physically build a structure with special blocks that input their arrangement to the computer.

Ultimately, it seems that the fastest input will require a certain amount of human-like intelligence on the part of the computer. The architect will describe what is required in general terms and the computer will 'work up' the drawing just as a good technician does at present. The drive to produce an intelligent computer is going on all over the world and such machines are referred to as the 'fifth generation'[158]. The Japanese may be the most advanced in this field and the Japanese government is funding a body known as the Institute of New-Generation Computer Technology[159]. The Institute acts as a research and coordination body and also commissions projects from commercial companies. It is scheduled to produce a fifth-generation machine by March 1992.

While on this topic, it is interesting to compare the current state of technology with the human brain. The brain is estimated to contain ten thousand million nerve cells, or neurons[160], and this can be compared with one of the most advanced experimental integrated circuits which contains the equivalent of about 1½ million discrete elements on a 12 mm square substrate[161]. In these terms, therefore, a moderately large machine could have the same complexity as the brain. However, the problem is not that simple for two reasons. The first is that each neuron does a certain amount of simple processing and so cannot be considered the equivalent of a single transistor or diode. The second is that the neurons are connected in parallel, with millions of pathways being active at once, whereas conventional computer design follows a single line of processing. Thus there must be a large increase in the number of elements per circuit and more importantly there must be a radical redesign of circuit functioning before a reasonable model of the brain can be produced[162]. Against this must be put the fact that the brain works by the chemical propagation of impulses while a digital computer works by electricity and the difference in speed is three million to one.

If computing goes on developing as it has been doing in the last thirty years, the future is one of fast and accelerating change. It seems inevitable that architects will be caught up in this flood, but whether the prospect is found to be exhilarating or terrifying, and whether an attempt is made to cope at all, must depend on the individual.

Glossary

acoustic coupler A form of *modem*.

address A location in a computer's memory.

algorithm A set of rules for solving a problem.

analogue computer A computer that uses a continously-variable quantity (usually electrical voltage) to represent numbers within itself.

APL (A Programming Language) A *high-level language* that makes extensive use of mathematical notation.

array A set of items of data arranged consecutively in computer memory.

array processor A type of computer incorporating a number of separate computational elements working simultaneously.

artificial intelligence A mode of working in which a computer mimics human intelligence.

ASCII (American Standard Code for Information Interchange) The most popular method of representing alphanumeric *characters* as numbers for computer storage.

assembly language A computer language that is very close to the basic instruction set of the machine; it is efficient but difficult for humans to use.

asynchronous transmission A mode of data transmission in which there is no fixed time between the items of data sent, as opposed to *synchronous transmission*.

background job A program in a computer that has low priority when competing for resources.

backing store A memory of large size, but low retrieval time compared with *main memory*; usually refers to *disc storage*.

backup The process of taking a copy of data in case of loss.

baseband transmission Transmission in which only one signal may be present in a channel at any time, as opposed to *broadband transmission*.

BASIC (Beginner's All-Purpose Symbolic Instruction Code) A *high-level language* that is easy to use and fairly powerful; the leading language on personal computers.

batch processing A system in which the user submits a program and returns later for the results, as opposed to *interactive operation*.

baud A measure of transmission speed; in most cases, 100 baud is equivalent to 10 *characters* per second.

benchmark A program used to measure the performance of a computer.

binary Having two possible states; the concept is appropriate to electrical circuitry as current may be flowing or not flowing.

bit (binary digit) The basic unit of information in a computer. A number of bits may be taken together to form a *character*.

bootstrap A process in which the first few lines of a program read in the rest. Normally used to start up a computer.

broadband transmission Transmission in which several signals may be present in a channel at any time, as opposed to *baseband transmission*.

bubble memory A form of solid-state *mass memory* in which information is stored by the arrangement of magnetised regions within a medium.

buffer store A store that holds data between one process and another, typically between a program and transmission to or from *mass memory*.

bug A fault in a computer program, or occasionally in the machinery, hence **debugging** the process of removing bugs.

byte see *character*.

C A *high-level language* often associated with the UNIX *operating system*.

CAAD Abbreviation of Computer-Aided Architectural Design.

CAD Abbreviation of Computer-Aided Design.

CADD Abbreviation of Computer-Aided Design and Draughting.

cartridge see *disc storage*.

Central Processing Unit (CPU) The part of the computer system that carries out the calculations and coordinates the other parts.

chain printer A printer on which the separate characters are on a continuous belt which is passed under a striking head.

character A single symbol such as a letter of the alphabet. It is usually represented within the computer by a group of eight *bits*, when it is also termed a *byte*.

check digit A number or letter added to an identification code in order to check its validity.

chip see *integrated circuit*.

clock A circuit that emits pulses at fixed intervals in order to control a process.

COBOL (COmmon Business Orientated Language) The most commonly used *high-level language*. It is orientated towards commercial applications.

compiler A computer program that transforms a program written in a *high-level language* into elementary machine instructions.

console On a large computer, the device used by the supervisor to give instructions to the machine.

continuous stationery Pages of printer paper which are joined into a continuous fan-folded strip.

CP/M (Control Program/Microcomputers) A popular *operating system* for personal computers.

cursor A visual indication on screen giving the currently referenced position. It usually takes the form of a cross in graphical applications and a flashing character in textual applications.

daisywheel printer A printer in which the characters are held in a ring which is rotated to bring the selected character under a striking head.

database A *file* which contains information structured into a logical form.

database management system (DBMS) A program which manipulates a *database* in order to read or write information.

debugging see *bug*.

digital computer A computer that holds and manipulates numbers in an exact form. Normally as assemblies of *bits*.

digitiser, or **tablet** A data-input device which will generate coordinates when touched with a special pen or puck.

disc storage A form of *mass memory* in which information is held as magnetised spots on the surface of a disc. The physical arrangement may take several forms: a **disc pack** is a number of discs on a common spindle; a **cartridge disc** is a single disc permanently enclosed in a rigid case; a **floppy disc** is relatively small, flexible and permanently enclosed in a cardboard sleeve; a **Winchester disc** is one or more discs permanently sealed within its housing.

dot-matrix printer A printer in which the characters are formed by driving forward some of a matrix of wires.

down A computer is said to be 'down' if it is out of action.

duplex transmission Data transmission which can take place in both directions at once, as opposed to *half-duplex*.

editor A computer program used to input and correct data or text.

execute To carry out a program within the computer.

expert system A computer program that mimics a human expert in a field.

facsimile transmission The process of scanning, transmitting and reproducing a pictorial image.

field A part of a *record* in a *database* which contains a single item of information.

fifth generation The next stage of computer development in which the aim is to achieve *artificial intelligence*.

file A distinct set of information stored in *mass memory*.

firmware Circuitry which has been programmed with instructions more commonly implemented as *software*.

floating-point number A number that can have a fractional part, as opposed to an *integer*.

floppy disc see *disc storage*.

flowchart A diagrammatic representation of the way in which a computer program is structured.

FORTRAN (FORmula TRANslation language) One of the earliest *high-level languages*, favoured in scientific and engineering applications.

half-duplex transmission A system in which data can travel in only one direction at a time, as opposed to *duplex transmission*.

hard copy Output printed on paper as distinct from that appearing on screen.

hard disc A form of disc storage in which the magnetic surface is rigid, as opposed to a *floppy disc*.

hardware Computers and computer circuitry.

head The device used to read or write information to a magnetic storage device.

hexadecimal Number representation to the base of sixteen.

high-level language A computer language that makes concessions to the programmer and is thus easier to use. It requires a *compiler* or *interpreter* to convert it into a form that the machine can understand.

host computer A central computer that provides information to one or more smaller computers linked to it.

icon A symbol representing a concept, used for giving instructions to a computer program.

IGES (Initial Graphic Exchange Specification) The most developed interface standard for the transmission of drawings in alphanumeric form.

ink jet printer A printer which forms characters by firing drops of ink at the paper.

integer A number that cannot have a fractional part, as opposed to a *floating-point number*.

integrated circuit An electronic circuit a few millimetres square constructed on a single wafer of silicon and embedded in plastic for protection. **Large Scale Integration (LSI)** describes such a circuit containing the equivalent of a few thousand elements. **Very Large Scale Integration (VLSI)** describes a circuit that may contain over a million elements.

interactive operation A system in which the computer and the user carry on a dialogue.

interpreter A computer program performing a similar function to a *compiler*, but without producing the intermediate stage of *object code*. It is however slower than a compiler in execution.

interrupt A diversion in the execution of a computer program caused by an external source.

I/O Abbreviation of Input/Output. These processes are regarded separately from the process of calculation because they are so much slower.

IT Abbreviation of Information Technology.

Job Control Language (JCL) A computer language which allows the user to give instructions to an *operating system* regarding the execution of a job.

joystick A data input device that is used for identifying points on a screen.

K The symbol for the number 1024 (2^{10}). Computers find it easier to work in such units as they are a power of two and so can be conveniently expressed in binary form.

Kilobyte (Kb) A group of 1024 *bytes* or *characters*.

Local Area Network (LAN) A system of passing information at high speed between computers. A dedicated wiring system is involved.

lightpen A data input device used to indicate points on a screen.

line printer A high-speed printer in which a complete line of text is set up then printed as a single operation.

LISP (LISt Processing language) A *high-level language* often used in *artificial intelligence* work.

LSI see *integrated circuit*.

machine code The basic language of a computer.

macro A group of commands which is named and can be invoked by giving that name.

magnetic disc see *disc storage*.

magnetic tape A form of *mass memory* in which information is held as magnetised spots on the surface of a tape. It is cheaper and more universal, but slower than *disc storage*.

mainframe A large computer.

main memory The memory of the computer itself, as distinct from *mass memory*.

Megabyte (Mb) A group of 1,048,576 (2^{20}) *bytes* or *characters*.

menu A list of alternatives from which a user can choose.

microcomputer A computer based on a *microprocessor*.

microprocessor A complete *central processing unit* constructed on a single *integrated circuit*.

microsecond (μs) One millionth of a second.

millisecond (ms) One thousandth of a second.

minicomputer A computer in which the *central processing unit* is constructed of discrete elements, but which is not physically large and does not require a controlled environment.

modem (modulator/demodulator) A device that converts electronic signals from a computer into acoustic signals for transmission by telephone and vice-versa.

mouse A device that inputs positional information into a computer as it is moved across a flat surface.

MTBF (Mean Time Between Failures) The average time that a device will operate between breakdowns. The most common measure of reliability.

multiplexor A communications control device that can pass information to a number of machines from data coming along a single channel. It does this by rapidly switching between the machines.

multiprogramming The system in which a number of programs exist in a computer's memory at once and whose execution proceeds in turn or as resources become available.

nanosecond (ns) A thousand-millionth of a second. Most *central processing unit* activities are measured in this unit.

non-volatile memory A form of memory in which information is not lost when power is removed.

object code The *machine code* output from a *compiler*.

OCR (Optical Character Recognition) A data input system in which the computer recognises the shapes of the letters and digits.

octal Number representation to the base of eight.

off-line An activity that does not involve the computer at the time of taking place, but did at some earlier time or will at some later time.

on-line An activity that directly involves the computer.

operating system The program that supervises the activity of a computer.

optical disc A form of *mass memory* in which information is stored as spots of differing light reflectivity on the surface of a disc.

package A computer program that can carry out a range of related functions in a certain application area.

parallel processing Computation in which a number of separate units are working simultaneously.

parameter An item of information which is passed to a computer program or *subroutine*.

parity bit A *bit* which is added to the end of a *character* to reflect whether it is odd or even as a form of checking.

PASCAL One of the more modern *high-level languages*.

peripheral device or **peripheral** A machine that is connected to the computer and is normally used to input data or output results.

pixel An element of a picture that is made up of a matrix of elements.

PL/1 (Programming Language 1) A *high-level language* that was designed for use in both scientific and commercial areas.

plotter An output device that produces drawings from computer instructions. The most popular type is the **drum plotter** in which the paper runs over a revolving drum and lines are drawn by combining motions of the drum and the pen. The **flatbed plotter** is more accurate but more expensive. In this type the paper lies on a flat surface and the pen alone moves. **Electrostatic plotters** use a row of needles to deposit a charge on the paper which attract ink particles to build up the picture.

plug-compatible Describes a device that will connect directly to a specified computer without modification.

port A socket on a computer or *peripheral* through which data is passed.

PROM (Programmable Read-Only Memory) A type of *ROM* which can be reset by an engineer when necessary.

punched card A system of representing data as holes in cards. Normally 80 columns and 12 rows are present. Now virtually obsolete.

RAM (Random Access Memory) The system used for the *main memory* of the computer. Data can be read or written at any point in a constant time.

raster screen A screen where the image is built up by repeatedly placing dots in parallel lines.

real time Actual elapsed time. A computer is operating in real-time mode, for example, if it is controlling traffic lights.

record A set of data items (*fields*) in a *database* that relate to a particular entity.

register A high-speed storage location within the computer which contains a number for use in computation.

relational database A type of *database* in which information is stored as a set of tables which are related to each other through common *fields*.

remote Communicating at a distance, as when a *peripheral* is placed at a significant distance from the computer.

report generator A program that takes the raw data in a *file* and outputs it as a document.

response time The time a computer takes to respond to a query.

ROM (Read Only Memory) A type of *main memory* which always contains a fixed set of information.

RS232C The most popular type of interface port. It has a defined shape and 25 pins.

scroll The facility in a *visual display unit* of being able to bring back text that has disappeared off the top of the screen.

simplex Equivalent to *half-duplex*.

simulation The process of building a mathematical model of a process within the computer and observing the results.

software Computer programs.

source code The typed version of a program in a *high-level language*.

spreadsheet A computer program that manipulates a matrix of data where the elements may be interconnected.

spooling (Simultaneous Peripheral Operation On-Line) The ability that larger computers have of performing input or output via *peripherals* while at the same time leaving the *central processing unit* free to perform calculations.

storage screen A type of screen in which the drawing is stored on the inside face of the screen.

streamer tape A type of high-speed *magnetic tape* unit.

subroutine or **subprogram** A distinct block of computer instructions to carry out a certain process. Reference can be made to the block whenever the process has to be carried out, thus avoiding duplication.

subscript A number referencing a particular element of an *array*.

synchronous transmission A mode of data transmission in which the items of data are sent at a fixed rate, as opposed to *asynchronous transmission*.

tablet see *digitiser*.

teleprinter A typewriter-like device used to communicate with a computer. A form of *terminal*.

terminal A device used by humans to communicate with a computer; most often a *visual display unit* or *teleprinter* will be used.

time sharing A process in which the computer takes advantage of the comparatively slow response time of humans by switching its attention between many users. Ideally, each user has the impression that he has the sole use of the computer.

touch-sensitive screen A screen which when a point on its surface is touched, will transmit the coordinates of that point to the computer.

trackball A ball-like device which when turned by hand will transmit the amount of movement and its direction to the computer.

user-friendly A term describing a computer program or system that is easy to use.

vector refresh screen A screen in which the computer repeatedly redraws the lines making up the image.

Visual Display Unit (VDU) A screen equipped with a keyboard. It is used to input data into the computer and show the results.

virtual memory A system in which the computer uses its *mass memory* as an extension of its *main memory*. It makes possible very large and complex programs but can be slow owing to the much lower speed of mass memory.

VLSI see *integrated circuit*.

volatile memory A form of memory in which information will be lost when power is removed.

Winchester disc see *disc storage*.

word A group of *characters* which represent a single number. The most common organisation on *microcomputers* and *minicomputers* is to have two characters to a word, while mainframes use four *characters* to a word.

word processing A system used to create and modify text *files*.

Bibliography

Chapter 1 - Past and Present

1 **Sutherland, I E** 'SKETCHPAD: a man-machine graphical communication system' *AFIPS Conference Proc.* Detroit, USA (1963), pp 329-346

2 **Newman, W M** 'An experimental program for architectural design' *Computer Journal* (May 1966), pp 21-26

3 **Alexander, C** *Notes on the Synthesis of Form*, Harvard University Press (1964)

4 **Boston Architectural Centre** 'Architecture and the Computer' *Proc. 1st Boston Architectural Centre Conf.* Boston Architectural Centre, Boston, MA, USA (December 1964)

5 **Ministry of Public Building and Works** 'Computers in the Construction Industry' *Conference Proc.* MoPBW, London, UK (1967)

6 **Campion, D** *Computers in Architectural Design*, Elsevier, London, UK (1968)

7 **American Institute of Architects** *Emerging Techniques of Architectural Practice: Study by the Engineering Dept. Pennsylvania State Univ.*, AIA, Washington, DC, USA (April 1966)

8 **Ray-Jones, A** 'Computer developments in West Sussex' (two parts) *Architects' Journal* (February 1968) pp 421-426 and pp 489-498

9 **Ministry of Health** *Hospital Design Notes* (five parts), HMSO, London, UK (1964-1968)

10 **Davis, C** 'The Harness System' *Hospital Engineering*, (January 1974) pp 3-11

11 'DHSS launches yet another hospital system and drops Harness' *Architects' Journal* (September 1975) pp 449-450

12 **Stewart, C D and Lee F** 'Can a 54 year-old architectural practice find romance and happiness with an interactive computer system?' *Progressive Architecture*, (July 1977) pp 64-73

13 **Richens, P** 'The OXSYS system for the design of buildings' *CAD78 Conf. Proc.* IPC Science and Technology Press, Guildford, UK (1978) pp 633-645

14 *Britain 1985*, HMSO, London, UK (1984)

15 *New Technology Report* Centre for Construction Market Information, London, UK (1986)

Chapter 2 - Overview

16 **Rosenfield, S H (Ed)** *Architect's handbook of professional practice* American Institute of Architects (3 vols) (1985)

17 **Royal Institute of British Architects** *Architect's Appointment* RIBA, London, UK (1985)

18 **Stanton, P** *Pugin* Thames and Hudson, London, UK (1972)

19 **Auger, B** *The Architect and the Computer* Pall Mall, London, UK (1972)

20 **Kewney, G** 'Databases need intermediaries' *Computing* (January 1978) p 2

21 **Cooley, M J E** 'Contributions to discussion' *Computer-Aided Draughting Systems Conf. Proc.* IPC Science and Technology Press, Guildford, UK (1973) pp 91-93

22 **Health and Safety Executive** *Visual Display Units*, HMSO, London, UK (1983)

23 **Reynolds, R A** 'Human Performance in Interactive Graphics Operations' *Computer Journal* Vol 26 No 1 (1983) pp 93-94

24 **Shortcliffe, E H** *MYCIN: Computer-based Medical Consultations* Elsevier/North Holland, New York, NY, USA (1976)

25 Negroponte, N (Ed) *Computer Aids to Design and Architecture* Petrocelli Charter, New York, NY, USA (1976)

26 Applied Research of Cambridge Ltd *Fact sheet: GDS Productivity Data* ARC Ltd, Cambridge, UK (1983)

27 Rogers, H 'Capturing the Ordnance Survey's map archive' *Computer Graphics '84 Conf. Proc.*, Online Publications, Pinner, Middlesex, UK (1984)

Chapter 3 - Equipment

28 'Sierra arrives at last' *Computer News* (21 February 1985) p 3

29 Riley, J 'Prime expects its 10 mips by 1990' *Computer Weekly* (24 January 1985)

30 Tanenbaum, A S and Hagen, T 'Two programs, many UNIX Systems' *European UNIX Users Group Newsletter* Vol 3 No 4 (1983) pp 24-26

31 Exton Smith, H 'Dinosaurs adapt to fight extinction' *Computer Management* (November 1985) pp 48-49

32 Banks, M 'The floppy reigns' *Management Review* (October 1984) pp 37-42

33 Sambura, A, Gero, J and Cornell, D 'Icon-driven interfaces for drafting systems' *Computer Graphics '84 Conf. Proc.* Online Publications, Pinner, Middlesex, UK (1984)

34 Woods, J 'The OCR alternative' *Computer Systems* (January 1985) pp 61-62

35 'Cadscan cuts cost of digitising manual drawings' *Computing* (1 November 1985) pp 29-30

36 Savary, J 'Designing on the dotted line?' *PC Business World* (10 September 1985) pp 13-14

37 Deasington, R J *X 25 Explained* Ellis Horwood, Chichester, UK (1985)

38 Pye, C *Networking with Microcomputers* NCC Publications, Manchester, UK (1985) pp 13-14

39 Meadows, R and Parsons, A J *Microprocessors: Essentials Components and Systems*, Pitman (1983)

40 Piper, R 'IBM jumps ahead with the PC AT' *MicroDecision* (July 1985) pp 128-131

41 Evans, B 'Low cost computing: Implementation' *Architects' Journal* (March 1983) pp 53-62

42 Health and Safety Executive *Visual Display Units* HMSO, London, UK (1983)

Chapter 4 - Programs

43 Alcock, D *Illustrating Basic*, Cambridge University Press (1977)

44 Lehmkuhl, N K *FORTRAN 77: A Top-Down Approach* MacMillan (1983)

45 Oakley, S *LISP for Micros*, Butterworths (1984)

46 Blackburn, L and Taylor, M *Introduction to Operating Systems* Pitman (1985)

47 Bourne, S R *The UNIX System*, Addison-Wesley (1983)

48 *Software Selector* CICA/RIBA Publications (1986)

49 Cornwall, R 'The Building Computer Centre one year on' *Building* (22 March 1985) p 51

50 Djurdjeric, R 'Inconclusive debate leaves issue open' *Computer Weekly* (31 January 1985) p 26

51 Asija, P *How to Protect Computer Programs* Law Publishers, Allahabad, India (1984)

52 HMSO *Copyright (Computer Software) Amendment Act* HMSO, London, UK (1985)

53 'Councils vigilant on CAD flaws' *Building* (21/28 December 1984) p 8

Chapter 5 - Databases

54 Gillenson, M L *Database* Wiley (1985)

55 Mayne, A and Wood, M B *Introducing Relational Database* NCC Publications, Manchester, UK (1983)

56 Wood, M B *Introducing Computer Security* NCC Publications, Manchester, UK (1982)

57 Simons, G L *Privacy in the Computer Age* NCC Publications, Manchester, UK (1982)

58 'The RIBA Annual Report 1975' *RIBA Journal* (May 1976) pp 195-210

59 'Can Viewdata replace the printed word?' *New Civil Engineer* (22 November 1979) p 33

60 Wix, J and Beale, G 'Instant information for the building industry' *Building* (20 September 1985) pp 91-92

61 'Database for good practice' *Building* (11 January 1985) p 57

62 Atkinson, G 'Construction goes on-line' *Building* (9 March 1984) p 49

63 'Computer Generated Schedules' *RIBA Journal* (December 1984) pp 54-55

Chapter 6 - Computer-Aided Draughting

64 Coutts, N 'Graphics: potential uses and limitations' *Architects' Journal* (December 1984) pp 77-85
65 Bijl, A and Shawcross, G 'Housing site layout system' *Computer Aided Design* (June 1975) pp 2-10
66 Richens, P 'The OXSYS system for the design of buildings' *CAD78 Conf. Proc.* IPC Science and Technology Press, Guildford, UK (1978) pp 633-645
67 Bijl, A Stone, D and Rosenthal, D *Integrated CAAD Systems* University of Edinburgh, UK (1979)
68 *MEDALS* Computer-Aided Design Centre, Cambridge, UK (1975)
69 Dunn, N *CARBS - User Introduction Manual* Clwyd County Council, Wales, UK (1974)
70 Reynolds, R A 'A comparison of graphical information handling techniques in architectural practice' *CAD78 Conf. Proc.* IPC Science and Technology Press, Guildford, UK (1978) pp 62-74
71 Stevens, T 'Low Cost CAD' *RIBA Journal* (December 1984) pp 50-51
72 Sarson, R 'Micro in practice: a learning experience' *CADCAM International* (February 1985) p 24
73 *Computer Draughting in Construction* CICA, Cambridge, UK (1980)
74 *The Automation of Draughting Work' International Federation of User Associations* translated and published by CICA, Cambridge, UK (1981)
75 Taylor, S 'Computer aided plotting' *Construction* No 47 (Summer 1984) pp 28-31
76 *Initial Graphics Exchange Specification (IGES) Version 2.0* National Bureau of Standards, Washington, DC, USA (February 1983)
77 Jeffreys, S 'The Engineer's WordStar' *PC Business World* (29 October 1985) pp 16-17
78 'Boost for AutoCAD' *PC Business World* (29 October 1985)
79 Poulson, T 'CAD - how to choose and how to use it' *What's New in Building* (October 1985) pp 149-151

Chapter 7 - Visualisation

80 Griffiths, J G 'A bibliography of hidden-line and hidden-surface algorithms' *Computer-Aided Design* Vol 10 No 3 (May 1978) pp 203-206
81 Sutherland, I E, Sproull, R F and Schumacker, R A 'A characterization of ten hidden-surface algorithms' *Computer Survey* Vol 6 No 1 (March 1974)
82 Clark, A L 'Roughing it: realistic surface types and textures in solid modelling' *Computers in Mechanical Engineering* (March 1985) pp 12-16
83 Crow, F C 'Shadow algorithms for computer graphics' *Computer Graphics* Vol 11 No 2 (Summer 1977)
84 Campion, D and Robey, K G 'Perspective drawing by computer' *The Architectural Review* (November 1965) pp 380-386
85 Bensasson, S *Computer Programs for Building Perspectives* DOC Ltd, Cambridge, UK (1977)
86 Bin, H 'Inputting constructive solid geometry representations directly from 2D orthographic engineering drawings' *Computer-Aided Design* Vol 18 No 3 (April 1986) pp 147-155
87 Aish, R and Noakes, P 'Architecture without numbers - CAAD based on a 3D modelling system' *Computer-Aided Design* Vol 16 No 6 (November 1984) pp 321-328
88 Boyse, J W and Pickett, M S (Eds) *Solid Modelling by Computers: from Theory to Applications* Plenum Press, New York, NY, USA (1985)
89 Pipes, A 'Solid modellers: this year's style' *CADCAM International* (September 1985) pp 20-23
90 Bishop, A W 'Dimensioning and tolerancing in CAD systems' *Draughting and Design* (19 January 1984) p 7
91 Webster, C A G *CAPITOL Users' Manual* Graphicsaid, Cleveland, UK (February 1985)

Chapter 8 - Job Management

92 'Easing the Complexity of Large Contracts' *Building* (19 April 1985) p 75
93 Bennett, F L *Critical Path Precedence Networks* Van Nostrand-Reinhold, New York, NY, USA (1970)
94 'Building project management' *Computer Systems* (May 1985) pp 15-16
95 RIBA *Architect's Job Book* RIBA Publications, London, UK (1977)
96 Peters, G *Construction Project Management Using Small Computers* Morgan-Grampian (1985)
97 'Design and Management by Micro' *Building* (3 May 1985) p 53
98 Stafforth, C and Wolton, H (Eds) *Programs for Network Analysis*, NCC Publications, Manchester, UK (1974)

Chapter 9 - Design Aid Programs

 99 *Industrial Engineering: Facilities Planning, Design and Management Issue* Vol 17 No 5 (May 1985)
100 **Doidge, C W** *University Space Utilization* (Doctoral Thesis), University College, London, UK (1969)
101 **Simons, G L** *Expert Systems and Micros* NCC Publications, Manchester, UK (1985)
102 **Gero, J and Coyne, R** 'The place of expert systems in architecture' *CAD84 Proc. Sixth International Conf. on Computers in Design Engineering* Butterworth Scientific, Guildford, UK (1984) pp 522-528
103 **Lansdown, J** *Expert Systems: Their Impact on the Construction Industry* RIBA Conference Fund, London, UK (1982)
104 **Radford, A D and Gero, J S** 'Towards generative expert systems for architectural detailing' *Computer-Aided Design* Vol 17 No 9 (November 1985) pp 428-435
105 **Wagner, D M** *Expert Systems and the Construction Industry* CICA, Cambridge, UK (October 1984)
106 **Poole, T and Szymankiewicz, J** *Using Simulation to Solve Problems* McGraw-Hill (1977)
107 **Reynolds, R A** *Comparison of Six Simulation Systems* (Internal Report) Cusdin Burden and Howitt, London, UK (1972)
108 'Process planning problems? First, see how with SEE-WHY' *The Production Engineer* (December 1981) pp 14-16
109 **Fleming, J** *LIFT-A1: A Computer Simulation Design of Lift Installations in Buildings* ABACUS, Glasgow, UK (1973)
110 **Marmot, A and Gero, J S** 'Modelling elevator lobbies' *Building Science* Vol 9 No 4 (1974) pp 277-288
111 **Laing, L W W and Gentles, J** 'AIR-Q, A computer model for designing airport terminal buildings' *Simulation Council Proceedings* Vol 6 No 1 La Jolla, CA, USA (1978)
112 *Movement in Buildings Simulator* ARC Ltd, Cambridge, UK (1973)
113 **Allwood, R** 'Expert system shells' *Architects' Journal* (13 November 1985) pp 71-76
114 **Levin, P H** 'The use of graphs to decide the optimum layout of buildings' *Architects' Journal* (October 1964) pp 809-815
115 **Whitehead, B and Eldars, M Z** 'An approach to the optimum layout of single-storey buildings' *Architects' Journal* (June 1964) pp 1373-1380
116 **Armour, G C and Buffa, E S** 'A heuristical algorithm and simulation approach to relative location of facilities' *Management Science* (February 1963) pp 294
117 **Kruskal, J B** 'Nonmetric multidimensional scaling: a numerical approach' *Psychometrika* (June 1969) pp 1-27
118 **Archer, L B** 'Computers in building: planning accommodation for hospitals and the transportation problem' *Architects' Journal* (July 1963) pp 139-142
119 **Needham, R M** 'Application of the theory of clumps' *Mechanical Translation* Vol. 8 (1965)
120 **Gero, J S, Julian, W and Holmes, W N** *The Development of a System for Heuristical Optimization of Topological Layouts Interaction* University of Sydney, Australia (1973)

Chapter 10 - Environmental Analysis

121 **Department of the Environment** *The Building (First Amendment) Regulations* HMSO, London, UK (1978)
122 *CIBS Guide, Section Al - Environmental Criteria for Design* Chartered Institution of Building Services, London, UK (1978)
123 *ASHRAE Handbook of Fundamentals* American Society of Heating, Refrigeration and Air-Conditioning Engineers, New York, NY, USA (1972)
124 *BRE Digest No 108, Standard U-Values* Building Research Establishment (1975)
125 **Bensasson, S and Burgess, K** *Computer Programs for Daylighting in Buildings* DOC Ltd, Cambridge, UK (1978)
126 *IES Code: Recommendations for Lighting Building Interiors* Illuminating Engineering Society, London, UK (1968)
127 **ASHRAE Task Group on Energy Requirements for Heating and Cooling of Buildings** *Algorithms for building heat transfer subroutines* American Society of Heating, Refrigeration and Air-Conditioning Engineers, New York, NY, USA (1975)
128 **Burgess, K S** *Computer Programs for Energy in Buildings* DOC Ltd, Cambridge, UK (1979)
129 'Programs to save energy' *Building* (14 December 1984) p 49
130 'Saving by numbers' *Building Design: Energy and Insulation Supplement* (May 1985)

131 Gronhoug, A C *Computer Terminal Programs to Aid Economic Selection of Fuel for Heating Systems* Department of the Environment (1977)
132 Peckham, R J 'Shading evaluations with general three-dimensional models' *Computer-Aided Design* Vol 17 No 7 (September 1985) pp 305-310
133 Cox, G and Kumar, S *BRE Information Paper IP2/83: Computer Modelling of Fire* Building Research Establishment (1983)
134 Ruffle, S 'A microcomputer model of noise acoustics in open-plan office layouts' *CAD82 Conf. Proc.* Butterworths, Guildford, UK (1982) pp 556-561
135 *BSRIA Application Guide 1/81: Locating Fire Alarm Sounders for Audibility* Building Services Research and Information Assoc., Bracknell, UK (1981)
136 Edwards, N 'Considering concert acoustics in the shape of rooms' *Architectural Record* (August 1984) pp 133-137
137 Vincent, G and Peacock, J *The Automated Building* Architectural Press (1985)
138 Clarke, J A *Energy Simulation in Building Design* Adam Hilger Ltd, Bristol and Boston (1985)

Chapter 11 - Office Management

139 Kennedy, K 'Job costing a way to better profits' *Building* (5 October 1984) p 73
140 'Computer revolutionises building accounting' *Building* (9 November 1984) p 51
141 Sarvary, J 'Why Pegasus is a market leader' *PC Business World* (February 1985) pp 29-31
142 Derrick, J and Openheim, P *The Word Processing Handbook* Kegan Page Ltd (1984)
143 Cassidy, B 'Making the most of word processing' *Building* (8 February 1985) p 57
144 Newman, J 'Supercharged WordStar' *MicroDecision* (July 1985) pp 56-60
145 Hutzel, I 'PC-Based Presentation Graphics' *Computer Graphics World* (May 1985) pp 62-70
146 Day, N and Rees, O *Computer Spreadsheets* BBC Publications, London, UK (1985)
147 'Ahead of their time?' *PC Management* (May 1985) pp 19-23

Chapter 12 - The Future

148 Duncan, R *Critics' Gaffes* Futura Publications London, UK (1984)
149 *Evening News* London, UK (September 1933)
150 Clarke, A C *Profiles of the Future* Pan Books (1983)
151 Troll, J H 'The thinking of men and machines' *Atlantic Monthly* (July 1954) pp 62-65
152 May, M D 'Occam' *SIGPLAN Notices* (April 1983) p 69
153 Edwards, D 'All about X-Stream Services' *Telecomms Monthly* No 9 (September 1985) pp 14-16
154 Lewis, P 'A high fibre video diet' *The Times* London, UK (5 June 1986)
155 Curtis, T 'Dazzling duo put wafers in limelight' *Computer Weekly* (21 March 1985) p 19
156 Cook, R 'The light, fantastic optical storage disc' *PC Business World* (February 1985) pp 48-50
157 McGrath, R A 'Talking to your keyboard' *Computer Graphics World* (April 1986) pp 79-82
158 Simons, G L *Introducing Fifth Generation Computers* NCC Publications, Manchester, UK (1983)
159 Ostler, N 'The proof of Japan's pudding', *Computer News* (4 July 1985) p 26
160 Von Neumann, J *The Computer and the Brain* Yale University Press (1958)
161 'IBM makes experimental million-bit memory chip' *Perspective* IBM (UK), Basingstoke, UK (July 1984) p 7
162 Iliffe, J K *Advanced Computer Design* Prentice-Hall, NJ, USA (1985)

Index